Micro- and Ultrafiltration Performance Specifications Based on Microbial Removal

Awwa Research Foundation

About the Awwa Research Foundation

The Awwa Research Foundation (AwwaRF) is a member-supported, international, nonprofit organization that sponsors research to enable water utilities, public health agencies, and other professionals to provide safe and affordable drinking water to consumers.

The Foundation's mission is to advance the science of water to improve the quality of life. To achieve this mission, the Foundation sponsors studies on all aspects of drinking water, including supply and resources, treatment, monitoring and analysis, distribution, management, and health effects. Funding for research is provided primarily by subscription payments from approximately 1,000 utilities, consulting firms, and manufacturers in North America and abroad. Additional funding comes from collaborative partnerships with other national and international organizations, allowing for resources to be leveraged, expertise to be shared, and broad-based knowledge to be developed and disseminated. Government funding serves as a third source of research dollars.

From its headquarters in Denver, Colorado, the Foundation's staff directs and supports the efforts of more than 800 volunteers who serve on the board of trustees and various committees. These volunteers represent many facets of the water industry, and contribute their expertise to select and monitor research studies that benefit the entire drinking water community.

The results of research are disseminated through a number of channels, including reports, the Web site, conferences, and periodicals.

For subscribers, the Foundation serves as a cooperative program in which water suppliers unite to pool their resources. By applying Foundation research findings, these water suppliers can save substantial costs and stay on the leading edge of drinking water science and technology. Since its inception, AwwaRF has supplied the water community with more than $300 million in applied research.

More information about the Foundation and how to become a subscriber is available on the Web at **www.awwarf.org**.

Micro- and Ultrafiltration Performance Specifications Based on Microbial Removal

Prepared by:
Joseph G. Jacangelo
MWH, Lovettsville, VA 20180

Nancy L. Patania Brown
MWH, Tucson, AZ 85716

Alain Madec and **Kellogg Schwab**
The Johns Hopkins University, Baltimore, MD 21205

Debra Huffman
University of South Florida, St. Petersburg, FL 33701

Gary Amy
University of Colorado, Boulder, CO 80309

Chandra Mysore and **Jerome Leparc**
Veolia Water North America, Norcross, GA 30029

Annette Prescott
Thames Water, Reading, England

Sponsored by:
Awwa Research Foundation
6666 West Quincy Avenue, Denver, CO 80235-3098

Published by:

Awwa Research Foundation

American Water Works Association

IWA Publishing

DISCLAIMER

This study was funded by the Awwa Research Foundation (AwwaRF). AwwaRF assumes no responsibility for the content of the research study reported in this publication or for the opinions or statements of fact expressed in the report. The mention of trade names for commercial products does not represent or imply the approval or endorsement of AwwaRF. This report is presented solely for informational purposes.

CONTENTS

TABLES

FIGURES

FOREWORD

The Awwa Research Foundation (AwwaRF) is a nonprofit corporation that is dedicated to the implementation of a research effort to help utilities respond to regulatory requirements and traditional high-priority concerns of the industry. The research agenda is developed through a process of consultation with subscribers and drinking water professionals. Under the umbrella of a Strategic Research Plan, the Research Advisory Council prioritizes the suggested projects based upon current and future needs, applicability, and past work; the recommendations are forwarded to the Board of Trustees for final selection. The foundation also sponsors research projects through an unsolicited proposal process; the Collaborative Research, Research Applications, and Tailored Collaboration programs; and various joint research efforts with organizations such as the U.S. Environmental Protection Agency, the U.S. Bureau of Reclamation, and the Association of California Water Agencies.

This publication is a result of one of these sponsored studies, and it is hoped that its findings will be applied in communities throughout the world. The following report serves not only as a means of communicating the results of the water industry's centralized research program but also as a tool to enlist the further support of the nonmember utilities and individuals.

Projects are managed closely from their inception to the final report by the foundation's staff and large cadre of volunteers who willingly contribute their time and expertise. The foundation serves a planning and management function and awards contracts to other institutions such as water utilities, universities, and engineering firms. The funding for this research effort comes primarily from the Subscription Program, through which water utilities subscribe to the research program and make an annual payment proportionate to the volume of water they deliver and consultants and manufacturers subscribe based on their annual billings. The program offers a cost-effective and fair method for funding research in the public interest.

A broad spectrum of water supply issues is addressed by the foundation's research agenda: resources, treatment and operations, distribution and storage, water quality and analysis, toxicology, economics, and management. The ultimate purpose of the coordinated effort is to assist water suppliers to provide the highest possible quality of water economically and reliably. The true benefits are realized when the results are implemented at the utility level. The foundation's trustees are pleased to offer this publication as a contribution toward that end.

Walter J. Bishop
Chair, Board of Trustees
Awwa Research Foundation

Robert C. Renner, P.E.
Executive Director
Awwa Research Foundation

ACKNOWLEDGMENTS

The authors of this report are indebted to the following water utilities and individuals for their cooperation and participation in this project:

Atlanta Fulton County (GA) Water Resources Commission: Michael Leonard
City of Scottsdale (AZ) Water Campus: William Vernon
San Jose Water Company (CA): Andrew Gere

The advice and support of the Project Advisory Committee, including Mark Clark, Andrew Gere, Paul Gagliardo and Nilaksh Kothari, and the AwwaRF project officers Jason Allen and Misha Hasan are appreciated.

The authors also acknowledge the valuable assistance and participation of the following individuals:

Jinhee Bae, The Johns Hopkins University

Gretchen Watson, University of Colorado
Fred Pontius, University of Colorado

Tracy Berg, University of South Florida
Angela Gennaccaro, University of South Florida

Briony Grose, Thames Water Utilities
Graham Bateman, Thames Water Utilities
Jaime Massanet-Nicolau, Thames Water Utilities
Beth Bradshaw, Thames Water Utilities
David Holt, Thames Water Utilities

Yvan Poussade, Veolia Water North America

Thanks are extended to Zenon Environmental Inc., US Filter and Polymem SA for their contributions of membranes and pilot plants. HF Scientific provided a bench-top turbidimeter and US Filter Stranco provided two on-line particle counters for pilot testing in the United States. Warren Johnson of Memcor, Jean-Michel Espenan from Polymem SA, Pierre Cote from Zenon Environmental, Inc. and Bruce Bartley from NSF International are acknowledged for their insightful comments on the project and report.

EXECUTIVE SUMMARY

INTRODUCTION

Low pressure membrane filtration is being increasingly employed worldwide for treatment of water supplies, particularly for those for which risk of disease from microorganisms is an issue. That microfiltration (MF) and ultrafiltration (UF) are capable of removal of microorganisms has been well demonstrated in the literature. If membranes are considered strictly as physical strainers that are not impacted by water quality and membrane or microbial surface characteristics, then any microorganism larger than the largest exclusion size of the membrane will be removed, and those smaller will pass through. However, other physical and chemical factors can lead to the removal of microorganisms that are smaller than the pores of the membrane. The traditional criterion for determining whether a membrane was considered a UF or MF membrane is the membrane's pore size distribution or molecular weight cutoff (MWCO). It should be noted that there are considerable difficulties and imprecision associated with the measurement of MWCO and pore size. MF membranes generally are considered to have pore sizes ranging from 0.05 μm to 5 μm and UF membranes from 0.005 μm to 0.05 μm. However, in practice there is considerable overlap between where UF may be considered to end and MF to begin in terms of pore size. The actual classification of these membranes for product marketing is based primarily on the manufacturer or vendor. In these traditional definitions, microbial removal does not play a role in determining whether a membrane is classified as MF or UF. This point leads to confusion in the water community as to the classification of low-pressure membranes. There is often a view in the industry that UF removes viruses and MF does not.

Both UF and MF have the capability to remove viruses (and submicron bacteria); however, the extent of removal is based on a number of factors including the membrane and the organism, as well as water quality and operational conditions. If membranes are to be employed on a more widespread basis for microbial removal, then their classification should be based on their ability to remove microorganisms, not on their nominal pore size.

PROJECT GOALS AND OBJECTIVES

The overall goal of this project was to develop a systematic performance testing protocol and specification for microfiltration and ultrafiltration membranes with respect to removal of viral and submicron bacterial pathogens. The following specific objectives combined both fundamental and practical studies to reach the overall goal. The specific objectives were to:

1. Conduct performance testing at bench scale with a variety of membranes and appropriate viral and submicron bacterial species or surrogates seeded into clean feedwaters;
2. Conduct performance testing at pilot scale using natural waters of varying water qualities with membrane materials, configurations and natural or seeded microbes as directed by bench-scale studies;
3. Conduct performance testing at full scale using natural water to validate pilot-scale data;
4. Develop a peer-reviewed protocol and performance specification based on data developed in the bench-, pilot- and full-scale testing for this project.

Key issues included the use of microbial surrogates for enteric viruses, and the impact of membrane characteristics, membrane operation and water quality variables on microbial removal.

SUMMARY OF PROJECT RESULTS

Bench-Scale Testing

Bench-scale testing was performed on a variety of membranes differing in material composition, configuration and manufacturer-reported MWCOs or pore sizes. Membrane feedwater was comprised of laboratory deionized, buffered water. Challenge organisms included MS2 and PRD1 bacteriophage, as well as a variety of animal viruses including poliovirus, hepatitis A virus and feline calicivirus. The membranes were also challenged with non-biological fluorescent microspheres. Overall findings of the bench-scale testing conducted for this project were:

- MS2 bacteriophage was a good surrogate for the enteric viruses studied (poliovirus, hepatitis A virus and feline calicivirus). In simultaneous testing using MS2, hepatitis A virus, poliovirus and feline calicivirus, there was no statistical difference between the mean removal of MS2 and that of any of the enteric viruses under the conditions tested. Use of MS2 bacteriophage as a viral surrogate was included in the bench-scale membrane testing protocol developed during this project.
- MS2 bacteriophage removal was generally lower than that of PRD1 bacteriophage; MS2 bacteriophage was therefore the worst-case surrogate.
- Use of PRD1 bacteriophage was included in the bench-scale membrane testing protocol developed during this project. PRD1 bacteriophage are used in the protocol in conjunction with MS2 bacteriophage in the microbial challenge experiments to better define the range of microorganisms that could be removed by a particular low-pressure membrane.
- Fluorescent microspheres were not a good surrogate for viruses and were not included in the bench-scale membrane testing protocol developed during this project. Consistent trends were not observed between the removal of MS2 bacteriophage and 26-nm microspheres, or between the removal of PRD1 bacteriophage and 67-nm microspheres.
- Manufacturer-reported MWCO or pore size was not predictive of microbial removal performance. Bench-scale testing was performed using five hollow-fiber, modified-polysulfone membranes that varied only in manufacturer-reported pore size or MWCO (same manufacturer, same membrane material), ranging from 10 kD to 0.2 μm. While the "loosest" membrane (0.2 μm) provided the poorest removal of MS2 and PRD1 bacteriophage and poliovirus, the 100- and 300-kD membranes provided consistently higher removal of the three organisms than did the "tightest" membrane (10 kD).
- Specific flux did not affect microbial removal performance under the conditions tested.
- Membrane integrity was crucial for microbial removal. Even small holes in the membrane area dramatically reduced microbial removal.
- pH did not affect microbial removal within a range typical of water treatment applications (pH 5 to 9).
- Ionic strength did not affect microbial removal within a range typical of water treatment applications.

Pilot-Scale Testing

Pilot-scale testing was performed in the United States at Atlanta and Scottsdale, and in the United Kingdom at London. Both pressurized and submerged membrane systems were employed, and a variety of membrane materials and manufacturer-reported MWCOs or pore sizes (0.1-µm PVDF in the US, and 0.035-µm PP and 150-kD PES/PVP in the UK). Membrane feedwaters included raw and filtered water from municipal drinking water treatment plants (Atlanta and London) and secondary effluent from a wastewater treatment plant (Scottsdale). Challenge organisms for the pilot-scale testing included MS2 and PRD1 bacteriophage as viral surrogates, and *Pseudomonas diminuta* bacteria. Pilot testing included initial determination of microbial removal under clean water conditions, which were the worst-case conditions for microbial removal (clean water, no deposition layer or fouling on the membrane). Following clean water testing at each pilot-testing location, microbial removal was evaluated under conditions of deposition layer formation or hydraulically-reversible fouling, followed by a longer-term evaluation of microbial removal under conditions of hydraulically-irreversible fouling.

Deposition Layer Formation

- In general, deposition layer formation as characterized by the decline in specific flux over the 4-hour experiments varied depending upon the membrane and feedwater employed.
- The natural raw waters tested had sufficiently high particulate and organic quality that deposition layer formation did not have consistent effects on microbial removal. Deposition layer formation consistently improved microbial removal only when secondary wastewater effluent or other high-TSS feedwater (TSS of approximately 5 mg/L or higher) was used as the membrane feedwater.
- In Scottsdale using secondary effluent as feedwater, the presence of a deposition layer increased PRD1 bacteriophage removal by a maximum of 3.1 logs, and increased MS2 bacteriophage removal by up to 1.6 logs compared to clean water conditions.
- In Atlanta, deposition layer formation increased PRD1 removal by a maximum of 1.8 logs and MS2 bacteriophage removal by a maximum of 1.3 logs with raw water; however, these increases were not consistent over all experimental conditions.
- In London, removal of both MS2 and PRD1 bacteriophage under clean water conditions was sufficiently high that consistent improvement with deposition layer formation was not observed.
- Increases in PRD1 bacteriophage removal with deposition layer formation compared to clean water conditions were consistently higher than those of MS2 bacteriophage removal in pilot testing at Atlanta and Scottsdale. This observation is consistent with the relative sizes of the two microorganisms, with PRD1 bacteriophage having the larger diameter (0.070 µm) compared to MS2 bacteriophage (0.024 µm).
- In summary, these experiments showed that less than 2 logs additional removal of virus could be achieved by a deposition layer on the membrane surface. As a practical matter for utilities, deposition layer should not be relied upon to provide consistent additional virus removal. The deposition layer is formed gradually after backwashing. As such, at the start of a filtration cycle immediately after backwash, the removal of virus achieved is equivalent to the “clean water” removal.

Membrane Fouling

- Membrane fouling had a much greater impact on microbial removal than did deposition layer formation under the conditions tested at all pilot plant locations. Declines in specific flux were greater during the longer-term membrane fouling experiments compared to the deposition layer formation experiments, and removal of PRD1 bacteriophage was consistently improved with membrane fouling compared to clean water conditions. Removal of the smaller-diameter MS2 bacteriophage was not as consistently improved with membrane fouling.
- Flux declines ranged from 30 to 90% for all but two of the 11 fouling experiments performed for this study.
- In all tests at Atlanta, membrane fouling led to increased removal of PRD1 bacteriophage, regardless of feedwater (raw or filtered) and membrane type (pressurized or submerged membranes). At the end of the experiment, maximum additional PRD1 bacteriophage removal compared to clean water conditions ranged from 1.6 to 4.9 logs at Atlanta. Additional removal of MS2 bacteriophage was not as consistent as that of PRD1, ranging from 0.0 to 0.9 logs for raw and filtered water with the submerged membrane and for filtered water with the pressurized membrane, but reached 5.0 logs for the pressurized membrane using raw water.
- In Scottsdale pilot testing using secondary effluent as feedwater, maximum flux decline over the 2-week experiments ranged from 30% for the submerged PVDF membrane, producing a maximum net PRD1 bacteriophage removal of 4.1 logs, to 90% for the pressurized PVDF membrane, which produced a maximum net PRD1 removal of 5.4 logs and a maximum net MS2 bacteriophage removal of 4.4 logs.
- In London, removal of both MS2 and PRD1 bacteriophage under clean water conditions was sufficiently high that consistent improvement in the fouling experiments was not observed.

Full-Scale Testing

Field testing comparing pilot-scale and full-scale membrane performance was performed in San Jose, California. At the 5-mgd full-scale membrane plant, raw surface water is subjected to pressure-driven membrane filtration using polypropylene membranes with a manufacturer-reported pore size of 0.2 μm. The full-scale membrane system comprises six membrane racks, each containing 90 membrane modules. The pressurized pilot unit was operated side-by-side with the full-scale membrane plant. Three polypropylene membrane modules were evaluated. One brand-new module was installed and operated in the pilot plant. One brand-new module was installed in one rack of the full-scale plant. The third module evaluated was one of the modules that had been in service in the full-scale plant since plant start-up approximately five years previously. In order to evaluate the microbial removal performance of the three modules, each was installed in the pilot unit where microbial seeding could be performed. Upon completion of the microbial seeding, each of the two modules from the full-scale plant was replaced in the full-scale membrane rack to resume operation until the next microbial challenge, when each was again removed from the full-scale rack and installed in the pilot plant.

- Side-by-side testing indicated that the hydraulic performance of the pilot-scale membrane module was comparable to that of the full-scale membrane rack. Specific flux was considerably higher in the pilot-scale membrane module initially, because it was brand new and was being compared to the full-scale membrane rack that contained 89 mature membrane modules and one brand-new module. However, after approximately two weeks of operation using the natural raw water as feedwater, the specific flux of the pilot membrane module stabilized no more than approximately 25% higher than that of the full-scale membrane rack. Flux recovery following chemical cleaning was consistent between the pilot- and full-scale membrane modules.
- Using both clean water (dechlorinated tap water) and raw water as feedwater, microbial removal performance of the brand-new membranes from the pilot- and full-scale membrane systems was comparable. Using clean water, neither of the two brand-new membrane modules, whether from the pilot- or the full-scale system, removed MS2 or PRD1 bacteriophage to any extent (maximum removal of 0.2 log). However, the mature membrane module removed 0.4 logs of MS2 bacteriophage and 2.2 logs of PRD1 bacteriophage. Thus, although the membranes were identical in every way except their history of use, PRD1 bacteriophage removal by the mature membrane was substantially higher that that of the brand-new membranes.
- When the membranes were operated side-by-side using natural feedwater, again there was little difference in MS2 bacteriophage removal between the pilot- and full-scale membrane modules, or between the brand-new and mature membranes. For PRD1 bacteriophage, there was little difference in removal between the brand-new pilot- and full-scale membrane modules. However, removal of PRD1 by the mature membrane module was higher by 1.0 to 2.0 logs than PRD1 bacteriophage removal by either of the two brand-new membranes.
- Thus, the key finding of the full-scale testing was that there were substantial differences in microbial removal performance between mature and brand-new membranes, while there was little difference in the microbial removal performance of pilot- versus full-scale membranes. New membranes produced substantially lower removal of PRD1 bacteriophage than mature membranes at similar specific fluxes. There was little removal of MS2 by any of the membranes, whether brand new or mature, or whether pilot scale or full scale.

Comparison of Bench-, Pilot- and Full-Scale Results

- Microbial removal performance was comparable at bench- and pilot-scale, whether under clean water conditions or in the presence of a deposition layer.
- Hydraulic and microbial removal performance was comparable in pilot-scale and full-scale membrane operation.

BENCH-SCALE PROTOCOL FOR CHARACTERIZING MEMBRANES USING MICROBIAL PROFILING

The overall goal of this project was to develop a performance testing protocol and specification for low pressure membranes with respect to removal of viral and submicron bacterial pathogens. An important task to meet this goal was to develop a peer-reviewed protocol and performance specification; this was followed by a second task which was to conduct performance testing at bench scale with a variety of different membranes using a clean water feed suspension of organisms. For pilot studies, there are various "accepted" protocols that are used for microbial challenge testing of membranes. However, there is currently no generally accepted methodology for microbial removal bench-scale testing of membranes used for drinking water.

The protocol that was developed in this study is presented in Chapter 3. As noted, it was designed for the application of sound science combined with ease of implementation by most qualified laboratories. To promote its use, it was developed with the idea that it would be employed under the Environmental Technology Verification (ETV) Program managed by the National Sanitation Foundation (NSF) for the US Environmental Protection Agency (EPA). As such, it has been adopted as the bench-scale option to perform microbial challenge studies for membranes (NSF International and USEPA, 2002). Additionally, the protocol was written with the intent to be employed as a bench-scale option described in the Guidance Manual for Membrane Filtration (USEPA 2003b) to the Long-Term 2 Enhanced Surface Water Treatment Rule. The draft Guidance Manual currently describes in detail how full-scale membrane modules should be challenged with microorganisms. However, it does not stipulate at any length the manner in which bench microbial challenge studies should be performed. Rather, it states, "For the purposes of consistency, it is recommended that manufacturers or independent testing agencies that opt to subject a product line to challenge testing using small-scale modules utilize a protocol that has been accepted by a wide range of stakeholders. Such a protocol has been proposed for use under the National Sanitation Foundation (NSF) Environmental Technology Verification (ETV)." The latter is the protocol developed under this study.

MEMBRANE CLASSIFICATION

Using the protocol developed for this study, hollow fiber membranes varying in manufacturer-reported MWCO/pore size as well as membrane materials were evaluated for removal of selected microorganisms. Table ES.1 presents a proposed virus and bacterial classification for the membranes studied. The table shows the average removal of MS2 and PRD1 bacteriophage as well as their respective lower confidence interval value around the average at P = 0.01. Also shown is a proposed classification of the membranes in terms of virus and bacterial removal. The scale employed is based on using half log values associated with the lower value of the confidence interval. The virus rating employs the removal of MS2 bacteriophage; this organism was chosen because of its small size (24 μm) and good correlation with removal of the enteric viruses studied. For example, the mPS 10 kD membrane had a lower confidence interval value of 3.8 log for removal of MS2; thus, its designation would be v-3.5.

For bacteria, removal of *P. diminuta* would be employed as the chosen organism by which to determine the bacterial rating of the membrane. This organism was selected based on its small size (0.6 to 1 μm). However, if no challenge studies with the bacteria were conducted and the membrane removed either MS2 or PRD1 bacteriophage to the RSL of the organism, the lower

Table ES.1
Viral and bacterial ratings of hollow fiber membranes employed in bench-scale study using Low-Pressure Membrane Testing Unit

		MS2 bacteriophage		PRD1 bacteriophage			
Module designation	MWCO or pore size*	Average log removal	Lower confidence interval value at P = .01	Average log removal	Lower confidence interval value at P = .01	Virus rating	Bacterial rating
mPs 10 kD	10,000 daltons	3.8	3.8	5.0	4.6	v-3.5	nc
mPS 100 kD	100,000 daltons	>5.7	5.7	>5.9	5.7	v-5.5	b-5.5
mPs 300 kD	300,000 daltons	>5.5	5.5	>5.2	4.8	v-5.0	b-5.0
PVDF 0.035 µm	0.035 µm	3.8	3.7	4.5	4.2	v-3.5	nc
PVDF 0.1 µm	0.1 µm	0.34	0.29	0.53	0.49	v-0	6.5†
mPS 0.2 µm	0.2 µm	0.45	0.22	nd	nd	v-0	nc

* As reported by the manufacturer

† Bacterial rating based upon the lower confidence interval removal value obtained from challenge studies of the membrane with *P. diminuta* (6.8 log-not shown on this table)

nd = no data

nc = not classified

value of the confidence interval for the removal of virus would be employed for the rating. If both PRD1 and MS2 data sets are generated, the lower value of the confidence interval for the bacteriophage that was removed to the greatest extent would be employed. For example, for the PVDF 0.1 μm membrane, challenge studies with *P. diminuta* were conducted and the lower confidence interval value for the removal of the organism was 6.8 log. Thus, the bacterial rating for this membrane would be b-6.5. However, for the mPS 300 kD membrane, no bacterial challenge studies were performed. Consequently, the MS2 bacteriophage data would be employed and the bacterial rating would assume the same rating as the virus and become b-5.0.

CHAPTER 1
INTRODUCTION

BACKGROUND

One of the primary drivers of the use of membranes in drinking water treatment has been the increased emphasis over the past two decades on the removal of microorganisms. The Surface Water Treatment Rule (SWTR; USEPA 1989), which was promulgated in 1989, requires utilities in the United States to provide a 3-log removal/inactivation of *Giardia* and a 4-log removal/ inactivation of viruses. The Interim Enhanced SWTR (IESWTR; USEPA 1998), finalized in December 1998, and the subsequent Long-Term 1 Enhanced Surface Water Treatment Rule (LT1ESWTR; USEPA, 2002), finalized in January 2002, require utilities to reduce the turbidity in filtered water from 0.5 to 0.3 NTU in 95 percent of the samples collected each month in order to achieve credit for 2 logs of *Cryptosporidium* removal. Additionally, no turbidity measurement over 1 NTU is allowable under these rules. The Long-Term 2 Enhanced SWTR (LT2ESWTR; USEPA, 2003a), proposed in August 2003, would require up to 3 logs of treatment for *Cryptosporidium* removal in addition to the requirements of the SWTR, IESWTR and LT1ESWTR, depending on the source water concentration of *Cryptosporidium* and the type of filtration treatment provided in the plant. Under the LT2ESWTR, *Cryptosporidium* removal credit of 2.5 logs or more can be achieved through properly demonstrated, operated and monitored membrane filtration.

In addition to water quality regulations, a decrease in availability of adequate water resources, and an emphasis on water for reuse have made membrane processes more viable as treatment processes. As advancements are being made in membrane technology, capital and operation and maintenance costs continue to decline, further encouraging the use of membrane treatment techniques.

That microfiltration (MF) and ultrafiltration (UF) are capable of removal of microorganisms was demonstrated in the early literature (Jacangelo et al. 1989, Jacangelo et al. 1991, Olivieri et al. 1991, McGahey 1994, Yoo et al. 1995). A summary of microorganisms that may potentially be removed by low-pressure membrane filtration and their relative sizes is shown in Figure 1.1.

If membranes are considered strictly as physical strainers that are not impacted by water quality and membrane or microbial surface characteristics, then any microorganism larger than the largest exclusion size of the membrane will be removed, and those smaller will pass through. However, other physical/chemical factors play a role in the removal of microorganisms (Fane 1994, McGahey 1994, Jacangelo et al. 1995, Madaeni 1998). For example, particles can be retained on the membrane surface and form a deposition layer that progressively accumulates as the filtration cycle continues. This cake may act as another physical barrier to the organism. Adsorption also plays a role in microbial removal. Organic and/or inorganic materials in the feedwater may adhere to the surface or inner matrix of the membrane resulting in smaller "pores" and enhanced removal. Further, the microorganisms themselves can be adsorbed onto the membrane through electrostatic adsorption. All of these factors can lead to the removal of microorganisms that are smaller than the pores of the membranes. Further, Herath et al. (1998) reported that virus rejection can be determined under ideal conditions from the ratio of their diameter to pore diameter. Madaeni et al. (1995) showed that operational parameters also impact virus rejection. Using stirred batched membrane filtration cells, the investigators found that the removal of the poliovirus by a polyvinylidene fluoride MF membrane was impacted by transmembrane flux and shear force.

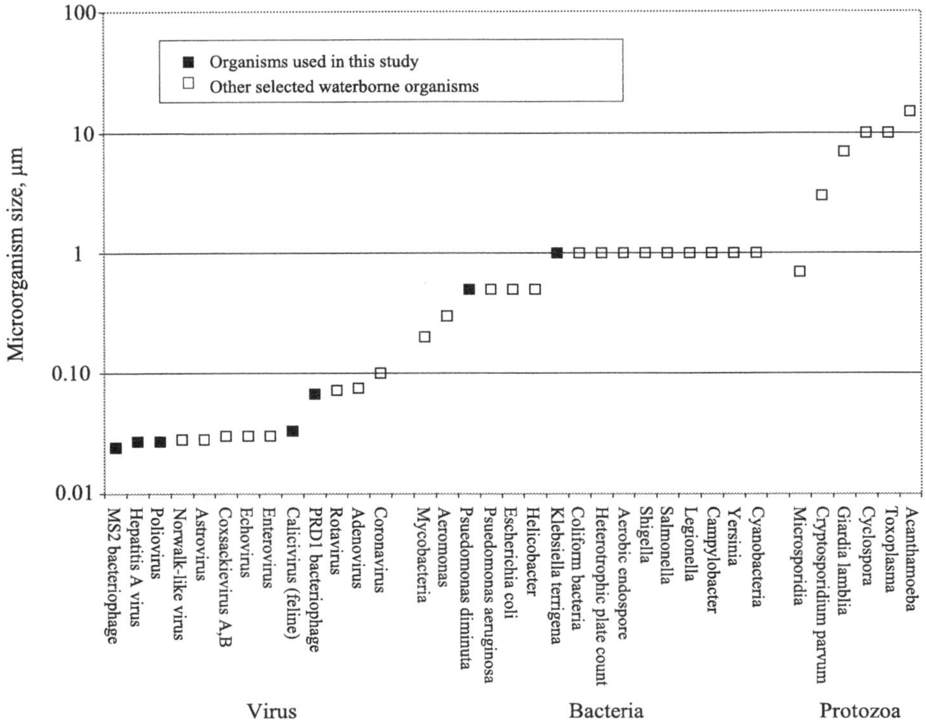

Figure 1.1 Selected waterborne organisms by approximate size

The highest viral rejections were observed under stirring conditions and lowest transmembrane pressure.

Although laboratory waters are useful to estimate the removal capabilities of membranes under conservative conditions, the extent of virus removal by low pressure membranes will vary in natural waters. In one of the earliest studies on microbial removal by low-pressure membranes, Olivieri et al. (1991) showed as much as four logs removal of male-specific bacterial virus from an activated sludge effluent by a polypropylene MF membrane with a 0.2-µm nominal pore size. That no virus was detected in the filtrate demonstrates an important difference between virus seeding studies and natural virus studies. In the former, the virus are normally prepared so that they are as monodispersed as possible (Jacangelo et al. 1991). Thus, the membranes are exposed to many single virions, which are not associated with particles or other microorganisms. In natural virus studies, to the contrary, the viruses can be to a great extent associated with particulate matter. By removing the particles, membranes can also remove substantial densities of viruses. Hence, viral seeding studies likely represent a conservative estimate of the true virus removal by MF and UF in natural waters.

The traditional criterion for determining whether a membrane was considered a UF or MF membrane is the membrane's pore size distribution or molecular weight cutoff (MWCO). It should be noted that there are considerable difficulties and imprecision associated with the measurement of MWCO and pore size. MF membranes generally are considered to have pore sizes ranging from 0.05 µm to 5 µm and UF membranes from 0.005 µm to 0.05 µm. However, in practice there is considerable overlap between where UF may be considered to end and MF to begin in terms of pore size. The actual classification of these membranes for product marketing is based primarily on the manufacturer or vendor. In these traditional definitions, microbial removal does not play a role in determining whether a membrane is classified as MF or UF. This point leads to confusion in the water community as to the classification of low-pressure membranes. There is often a view in the industry that UF removes viruses and MF does not.

The above discussion emphasizes that both UF and MF have the capability to remove viruses (and submicron bacteria); however, the extent of removal is based on a number of factors including the membrane and the organism, as well as water quality and operational conditions. If membranes are to be employed on a more widespread basis for microbial removal, then their classification should be based on their ability to remove microorganisms, not on their pore size distribution.

PROJECT GOALS AND OBJECTIVES

The overall goal of this project was to develop a systematic performance testing protocol and specification for microfiltration and ultrafiltration membranes with respect to removal of viral and submicron bacterial pathogens. The following specific objectives combined both fundamental and practical studies to reach the overall goal. The specific objectives were to:

1. Conduct performance testing at bench scale with a variety of membranes and appropriate viral and submicron bacterial species or surrogates seeded into clean feedwaters;
2. Conduct performance testing at pilot scale using natural waters of varying water qualities with membrane materials, configurations and natural or seeded microbes as directed by bench-scale studies;
3. Conduct performance testing at full scale using natural water to validate pilot-scale data;
4. Develop a peer-reviewed protocol and performance specification based on data developed in the bench-, pilot- and full-scale testing for this project.

PROJECT OVERVIEW

The project was undertaken by a team of researchers and water treatment practitioners from the following organizations:

- Montgomery Watson Harza
- Veolia Water
- Thames Water
- Johns Hopkins University (JHU), Bloomberg School of Hygiene and Public Health
- University of South Florida (USF), College of Marine Science
- University of Colorado at Boulder (UCB), Department of Civil, Environmental and Architectural Engineering
- Atlanta/Fulton County Water Resources Commission, USFilter/Khafra
- City of Scottsdale Water Campus
- San Jose Water Company

Hollow fiber membrane bench-scale testing was performed at JHU to examine microbial removal by a variety of membrane types and materials under clean water conditions, as well as the impact of factors such as pH and ionic strength, as well as operating parameters such as flux. Bench testing employed the traditional viral surrogates MS2 and PRD1 bacteriophage; the human enteric viruses poliovirus type 1 and hepatitis A virus; as well as bacteria including *Escherichia coli* and *Pseudomonas diminuta*. Parallel flat sheet membrane bench testing at UCB examined the effects of membrane type and material, pH, the presence of fouling constituents including Ca^{2+} and natural organic matter (NOM), and the compromising of membranes on removal of MS2 and PRD1 bacteriophage and abiotic surrogates (26- and 67-nm microspheres).

Pilot-scale testing was performed by Veolia Water using both pressurized and submerged (vacuum-driven) pilot units at the Atlanta/Fulton County Water Treatment Plant and the Scottsdale Water Campus. Pilot-scale testing using pressurized and submerged pilot units was also performed by Thames Water at the Kempton Park Advanced Water Treatment Works in London. These tests evaluated the membranes using raw and filtered water from each of the facilities, and examined the removal of *P. diminuta* and MS2 and PRD1 bacteriophage under conditions of clean water, deposition layer formation and fouling.

Full-scale testing was performed by Veolia Water at the City of San Jose Water Company to compare long-term performance of pilot- and full-scale membrane systems and examine the removal of MS2 and PRD1 bacteriophage by new and mature membrane modules.

All microbial propagation and analysis was provided by USF or JHU for the bench-scale testing, the pilot testing performed in the US, and the full-scale testing.

A bench-scale protocol was developed as a membrane performance specification for microbial removal. The protocol is based on microbial removal rather than pore size or MWCO. The protocol was developed based on the bench-scale testing performed at JHU, and was peer reviewed by a large panel of experts. A total of five drafts were prepared and extensively reviewed. One or more of the drafts were reviewed by:

- National Sanitation Foundation (NSF)
- NSF Environmental Technology Verification (ETV) Program Steering Committee
- USEPA
- USEPA contractors for development of the LT2ESWTR
- Selected USEPA State Primacy agencies
- AwwaRF Project Advisory Committee for this project

Additional drafts were posted on the NSF's web site to solicit comments, and emails were sent to over 600 stakeholders noting the availability of the drafts for review. As an outcome of these reviews, the document is referenced in the USEPA's Membrane Filtration Guidance Manual to the LT2ESWTR (USEPA 2003b) as an eligible protocol to be followed for bench-scale evaluation of membranes. Further, the protocol was adopted by the NSF's ETV Program for verification of membranes for microbial removal under this program.

CHAPTER 2
METHODS AND MATERIALS

BENCH-SCALE STUDIES

Hollow Fiber Membrane Studies

Methods and materials for hollow fiber membrane studies are described in Chapter 3, Bench-Scale Protocol.

Flat Sheet Membrane Studies

General Laboratory Methods

All experiments were performed at room temperature (approximately 20°C). Milli-Q water (NOM-free, deionized water additionally purified with granular activated carbon and a 0.2-μm filter; Millipore Water Purification System, Bedford, MA) was used in preparation of all solutions and suspensions. pH was measured with an Orion 250A pH meter and glass probe, calibrated with two standards (pH 7 and pH 4 or 10) prior to each analysis. The pH of each dilution water used was adjusted with sodium hydroxide. The ionic strength of each dilution water used was adjusted with sodium chloride.

Microsphere and Bacteriophage Characterization

Microspheres and bacteriophage were characterized in the lab according to their zeta potential and hydrophobicity. The zeta potential was measured on a *ZetaPlus* zeta potential analyzer (Brookhaven Instruments Corporation) at a constant ionic strength of 0.01 M, provided by sodium chloride. The relative hydrophobicity of the microspheres and bacteriophage was determined through a simple procedure that measured the adsorption of the microspheres and bacteriophage to a hydrophobic polyvinylidine fluoride (PVDF) membrane that was used in all of the experiments. Suspensions of approximately10^{10} spheres/mL or approximately 10^6 pfu/mL were prepared using pH 6.5 dilution water. The hydrophobic PVDF membrane was submerged in 200 mL of the suspension for 1 hour. The microsphere or bacteriophage concentrations before and after the membrane was submerged, were determined and used to calculate removal. All other characterization of the microspheres was done by the manufacturer. Other characteristics of the bacteriophage were derived from the literature.

Membrane Characterization

Membranes were characterized according to zeta potential, contact angle, roughness, pore size or molecular weight cut-off (MWCO), and thickness.

Contact Angle. Contact angle is an indicator of the hydrophobicity of a membrane. Contact angle was measured using a goniometer (Rame-Hat Inc., NRL, CA). The sessile drop method used was based on measuring the angle between the water droplet and the membrane surface. The membrane sample was cut into a 2×2 cm^2 specimen and then mounted onto a glass support. 2.0 μL of Milli-Q water was placed on the membrane. The contact angle was measured within 10 seconds of application of the water droplet using a goniometer (Cho, 1998).

Zeta (ζ) Potential. The ζ potential of a membrane indicates the surface charge over a range of pH values and is defined as the potential across a fluid shear plane at the surface. The ζ potential of the membranes was determined using a commercial streaming potential measurement cell (BI-EKA, Brookhaven Instruments Corp., Holtsville, NY) to determine the streaming potentials, and then estimating the ζ potential values with the Helmholtz-Smoluchowski equation (Shim et al., 2002; Cho, 1998).

Roughness. The relative roughness of the membranes was determined in two ways. Initially, the membranes were examined using scanning electron microscopy (SEM) to provide a relative, qualitative ranking. Then, the more precise relative roughness was determined using atomic force microscopy (AFM). The instrument used was a Digital Instruments MMAFM-2 in the JILA W.M. Keck Optical Measurement Laboratory at UCB. The instrument maps the roughness of the membrane surface with a very fine tip that tapped and thus probed the membrane surface. This tip was connected to a cantilever, and the deflection of the cantilever was recorded by a diode laser. Three-dimensional maps of the undulation of the cantilever were generated.

Pore Size/Molecular Weight Cut-Off. The manufacturers provided estimates of pore size (μm) or molecular weight cutoff (MWCO, kD) for the membranes. A relationship between pore size and MWCO was determined through the use of molecules of a known molecular weight and estimated molecular diameter. These values were plotted and a curve fitted, resulting in an equation that related pore size to MWCO (Table 2.1 and Equation 2.1). The equation was modified such that the diameter of the pore d_{pore} (Å) could be substituted for the molecular diameter if the MWCO (Daltons) is substituted for the molecular weight of the molecule (Baker, 2000 and Wright, 2002).

$$d_{pore} = 15.452*(MWCO*10^{-3})^{0.3371} \tag{2.1}$$

Hydraulic Testing. An uncompromised membrane was placed in a stirred-cell apparatus. The pressure tank was filled with water and connected to the system. With the valve open on the stirred cell, nitrogen was allowed to flow into the pressure tank, flooding the system and the stirred cell with water. When the stirred cell was full, the valve on the stirred cell was closed. The system pressure was increased to approximately 1 bar (15 psi). The volume of water that permeated the membrane in the stirred cell was collected in a graduated cylinder and measured for 30- to 60-second increments at least three times. The pressure was then decreased and the process repeated. The pressure was decreased in 0.07- to 0.14-bar (1- to 2-psi) increments until 0.27 bars (4 psi) was reached. Then it was decreased in 0.03- to 0.07-bar (0.5- to 1-psi) increments until 0.03 bars (0.5 psi) was reached. The membrane was then compromised with a hole. The above steps were repeated. The same procedure was performed until the membrane had been compromised by 5 holes.

Scanning Electron Microscopy. The SEM used was a JEOL JSM-6400 with a magnification of 10× to 300,000× and a resolution of 3.5 nm. The instrument was located at UCB, JILA W.M. Keck Optical Measurement Laboratory, which is operated jointly by UCB and the National Institute of Standards and Technology. SEM was used to characterize the membrane surface and thickness, and also to measure the size of holes made when compromising the membrane. Hole sizes and membrane thickness were calculated using an imaging software (Scion Image).

Table 2.1
Reference molecule molecular weights and diameters

	Molecular weight (kD)	Estimated molecular diameter (Å)*
Sucrose	0.34	11
Raffinose	0.59	13
Vitamin B12	1.36	17
Bacitracin	1.41	17
Insulin	5.7	27
Cytrochrome C	13.4	38
Myoglobine	17	40
α-Chymotrysinogene	25	46
Pepsin	35	50
Ovalbumin	43	56
Bovine albumin	67	64
Aldolase	142	82
c-Globulin	150	84

Source: Adapted from Baker 2000 and Wright 2002.
* 1 nm = 10 Å

Bench-Scale Experimentation

Sample Handling/Storage. For each stirred-cell bacteriophage challenge, all glassware used was autoclaved and only sterile pipette tips were used. Additionally, before each stirred-cell bacteriophage challenge, the stirred-cell apparatus was disinfected. A 100-mg/L chlorine solution was allowed to stand in the stirred cell for 20 minutes. The solution was allowed to drain and then the procedure was repeated. The stirred cell was then filled with 3 M sodium thiosulfate solution that had been autoclaved, and allowed to stand for 20 minutes, and this procedure was repeated. Then the stirred cell was rinsed twice with the buffer to be used. Also, a new membrane was used for each stirred-cell bacteriophage challenge.

Stirred-Cell Filtration

Microsphere Protocol. A suspension of known concentration of microspheres was prepared from the microsphere stock using sterile dilution water at the desired pH (6.5 or 8.5), and ionic strength (0.01 M). In the case of microsphere challenges with fouling, calcium concentration was adjusted with calcium chloride. The membrane to be tested was placed in a 25-mm stirred cell, which had a volume of 10 mL, and rinsed twice with 10 mL of the buffer/dilution water. If the membrane was to be compromised, then compromising was done at this point. One mL of sample was evenly distributed on the wetted membrane, followed by 4 mL of buffer/dilution water. The membrane was then rinsed with another 5 mL of buffer/dilution water. All filtrate was collected and the volume was measured.

In all stirred-cell microsphere challenges, only one of the two microspheres was present. No microsphere and/or bacteriophage cocktails were used. Challenges using microspheres and bacteriophage were performed separately.

The microsphere concentration of the filtrate was determined by fluorescence spectroscopy. The excitation/emission wavelengths for 67-nm spheres and 26-nm spheres were 490/515 nm and 520/580 nm, respectively. The total filtrate volume collected was the minimum needed for triplicate measurements when the 25-mm stirred cell was used. Additional volume would have decreased the sensitivity of the analytical measurement. This protocol provided a count of the spheres that passed through the membrane into the filtrate. Knowing the concentration of microspheres in suspension before and after filtration, as well as the volume of suspension, allowed for calculation of microsphere removal.

The maximum challenge concentration of microspheres that was used allowed the highest removal ($\log_{10}$) to be calculated. This concentration did not exceed approximately 10% coverage of the surface area of the filter specimen to ensure that removal was not due to clumping or deposition layer formation. The concentration of 26-nm spheres used was approximately 10^{11}/mL and the concentration of 67-nm spheres was approximately 10^{10}/mL. The sensitivity limits of the microsphere assay are discussed below.

Bacteriophage Protocol. A suspension of known concentration of bacteriophage was prepared from the bacteriophage stock using sterile dilution water at the desired pH (6.5 or 8.5), and ionic strength (0.01 M). In the case of bacteriophage challenges with fouling, calcium was also adjusted. The membrane to be tested was placed in a 25-mm stirred cell, and rinsed twice with 10 mL of the buffer/dilution water. If the membrane was to be compromised, then that action was performed at this point. Five mL of sample was placed on the wetted membrane. The membrane was then rinsed with another 5 mL of buffer/dilution water. All filtrate was collected and the exact volume was noted.

In all stirred-cell bacteriophage challenges, only one of the two bacteriophage was present. No microsphere and/or bacteriophage cocktails were used. Challenges using microspheres and bacteriophage were performed separately.

The total filtrate volume collected was the minimum needed for bacteriophage analysis. The filtrate samples were packaged and shipped to USF for processing. This protocol provided a count of the bacteriophage that passed through the membrane into the filtrate. Knowing the concentration of bacteriophage in suspension before and after filtration, as well as the volume of suspension, allowed for calculation of bacteriophage removal ($\log_{10}$). The bacteriophage were applied at a concentration of 10^6 pfu/mL. The sensitivity limits of the bacteriophage assay are discussed below.

Microsphere and Bacteriophage Concentrations. Microsphere and bacteriophage loading concentrations were determined by calculating the number of microspheres or bacteriophage per membrane that were needed to achieve 10% coverage of the membrane. Ten percent was chosen as the maximum coverage to prevent clumping and caking from affecting microsphere or bacteriophage removal. Ten percent coverage was calculated by using the diameter of a microsphere or bacteriophage to calculate its projected area onto a flat surface (a circle with the same diameter as the sphere or organism). The effective membrane area, which was the part exposed to the microspheres or bacteriophage, was also calculated. The effective membrane area multiplied by 10% and divided by the projected area of a microsphere or bacteriophage provided the number of microspheres or bacteriophage that would cover 10% of the effective membrane surface.

Fouling. Microsphere and bacteriophage challenges with fouling constituents were conducted only for the two PVDF membranes at pH 6.5. The membrane was placed in the stirred cell and rinsed twice with 10 mL of fouling buffer (pH 6.5, I = 0.01 M, Ca^{++} = 100 mg/L). The flux was measured at 1 bar (15 psi) during the second rinse. The membrane was then rinsed three times with 20 mL of the fouling buffer with Suwanee River fulvic acid natural organic matter (NOM) added (5 mg/L as DOC). This equated to 0.03 mg DOC/cm^2 delivered to the membrane. The flux was measured for each 20 mL. The membrane was once again rinsed with 10 mL of the fouling buffer and membrane compromising was performed at this time if necessary. The microsphere or bacteriophage suspension sample was then applied to the wetted and fouled membrane, followed by a 5-mL rinse with the fouling buffer. All the filtrate was collected. The filtrate was then enumerated by either fluorescent spectroscopy or sent to USF for bacteriophage enumeration, depending on the nature of the filtrate.

Microsphere Enumeration. A series of different analytical protocols were evaluated as potential microsphere enumeration tools including: 1) epifluorescence microscopy, 2) field flow fractionation, 3) fluorometry, 4) optical particle counting, 5) UV absorbance, 6) laser turbidimeter, 7) turbidimeter, 8) coulter counter, 9) SEM, and 10) fluorescence spectrophotometry (Watson, 2002). Fluorescence spectrophotometry was finally selected because it was readily available, easy to use, and had an acceptable sensitivity limit. None of the other techniques evaluated had any advantages over fluorescent spectrophotometry. In order to overcome the inability of this technique to determine size distribution, samples could only contain one size of microsphere at a time.

Fluorescence spectrophotometry is a simple technique based on the detection of the emission of energy by an electron at an emission wavelength followed by excitation of the electron at an excitation wavelength. This method can accommodate a wide range of emission/excitation wavelengths. However there are a few disadvantages. The sample cell size limits the sensitivity of the instrument. Additionally, fluorescence spectrophotometry does not provide information on microsphere size distribution.

Microsphere enumeration was conducted by fluorescence spectrophotometry using a Hitachi F-3010 Fluorescence Spectrophotometer. The excitation/emission wavelengths for the 26- and 67-nm microspheres were 520/580 and 490/515 nm, respectively, as specified by the manufacturer. The calibration curves for both 26- and 67-nm spheres are shown in Table 2.2 and Figures 2.1 and 2.2. The data were linear at higher concentrations; at lower concentrations, the data in the tables were used.

In order to interpret stirred-cell data, a removal sensitivity limit (RSL) was estimated. The RSL was calculated for each fluorescent microsphere using the calibration curve and the feed concentration used in a stirred-cell challenge:

$$\text{removal sensitivity limit (RSL)} = -\log(N/N_0) \qquad (2.2)$$

where: N = final sample volume * lowest detected concentration from calibration curve
N_0 = # spheres applied = feed volume applied * feed concentration

The final sample volume was 10 mL and the feed volume was 1 mL for the 25-mm stirred cell.

Bacteriophage Enumeration. Bacteriophage enumeration was performed at the USF and is described below. The RSLs for PRD1 and MS2 bacteriophage were 5.2 and 4.9 logs, respectively.

Table 2.2
Calibration values for 26- and 67-nm microspheres

26-nm microspheres				67-nm microspheres			
	Fluorescence intensity				Fluorescence intensity		
#/mL	Mean	Standard deviation	#/mL (log_{10})	#/mL	Mean	Standard deviation	#/mL (log_{10})
1.26E+12	26.54	0.15	12.1	1.32E+12	421.99	5.69	12.1
1.26E+11	3.338	0.23	11.1	1.30E+11	45.762	0.424	11.1
1.26E+10	0.486	0.016	10.1	1.34E+10	4.627	0.048	10.1
1.26E+09	0.053	0.004	9.10	1.30E+09	0.417	0.005	9.11
1.26E+08	0.013	0.012	8.10	1.36E+08	0.057	0.021	8.13
1.26E+07	0.005	0.003	7.10	1.30E+07	0.026	0.009	7.11
1.26E+06	0.004	0.002	6.10	1.38E+06	-0.013	0.005	6.14
1.26E+05	0.003	0.004	5.10	1.24E+05	0.003	0.007	5.09
1.26E+04	-0.016	0.002	4.10	1.40E+04	-0.008	0.012	4.15

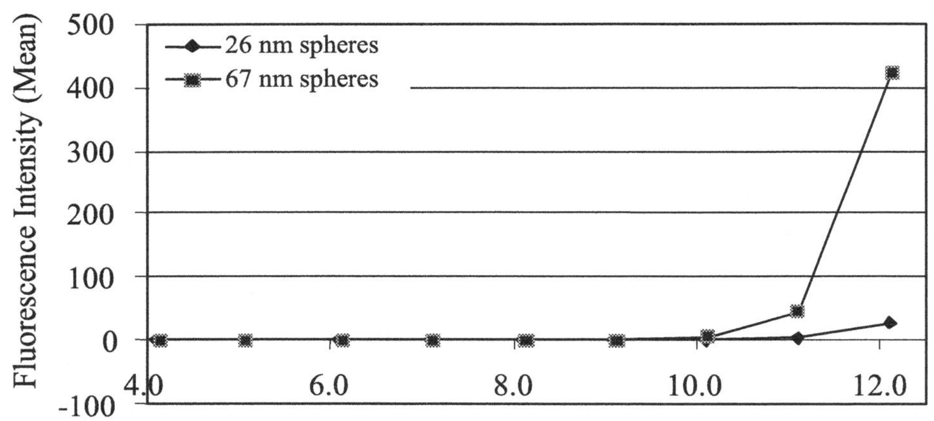

Figure 2.1 Calibration curve for 26- and 67-nm microspheres

Calculation of Microsphere and Bacteriophage Removal. Microsphere or bacteriophage removal was calculated using the following equation:

$$\text{removal} = -\log(N/N_0) \tag{2.3}$$

where N is the number of microspheres or bacteriophage in the filtrate and N_0 is the number of microspheres or bacteriophage in the applied suspension.

Membrane Integrity. Several methods of membrane compromising were evaluated (Watson, 2002) including: 1) needle or pin, 2) dental instruments, 3) surgical instruments;

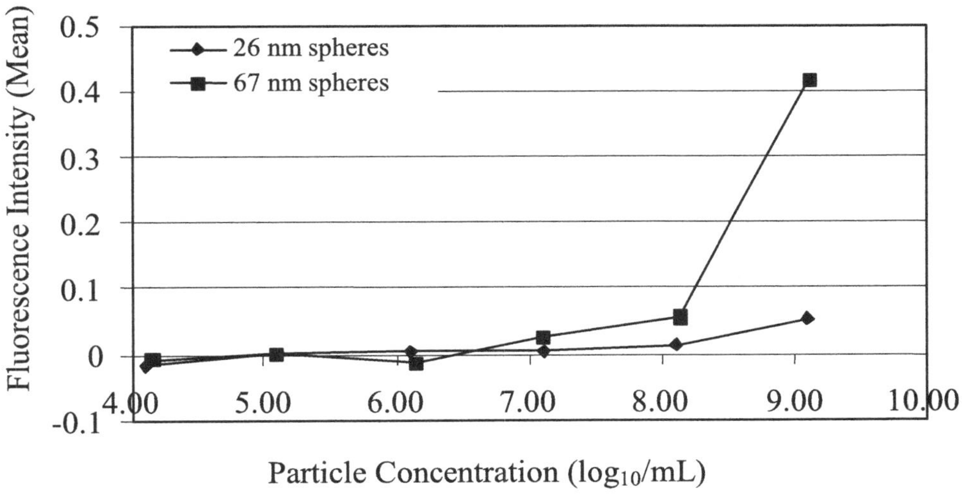

Figure 2.2 Calibration curve for 26- and 67-nm microspheres (detail of lower concentrations)

4) probes, 5) micro-dissecting probe, 6) microelectrodes, 7) thin wire, 8) fiber optic pulling, 9) lasers, 10) Excimer laser, and 11) focused ion beam milling. Initially, the goal was to find a technique that could consistently make 2.5- to 3.0-μm holes. Unfortunately this was not found to be possible. Consequently, a technique was sought that could make as small a hole as possible. It was found that the smallest hole that could be made in a flat-sheet membrane was 50 to 100 μm. The smaller holes in this range were made by an Excimer laser, while a sewing needle made the larger holes. It was found that the most cost-effective method of compromising was a sewing needle. The size advantage gained by using the Excimer laser was not great enough to justify the cost and inconvenience.

Bacteriophage, Microsphere and Membrane Characteristics

Bacteriophage Characteristics. The characteristics of the two bacteriophage used in this study are summarized in Table 2.3. PRD1 and MS2 bacteriophage both have an icosahedron shape, which is similar to the spherical shape of the microspheres (Herath et al., 1999). An icosahedron is a polyhedron with 20 faces. However, whereas the microspheres have essentially smooth surfaces, bacteriophage have rough surfaces. Since there is not a good method available for measuring the roughness of particles, this characteristic was ignored in analysis.

Because bacteriophage capsids are constructed from proteins which have both carboxylic acid and amino functional groups, the charge is a net negative charge that is caused by both carboxylic acid groups which can be negative and amino groups which can be positive. A net charge means there could be portions of the bacteriophage surface that exhibit different charges. The zeta potentials for each bacteriophage were determined at pH 6.5 and pH 8. Since there was little change over the pH range studied, the two values were averaged in Table 2.3. Nevertheless, PRD1 bacteriophage is more negative over the pH range studied than is MS2 bacteriophage. Additionally, Schijven and Hassanizadeh (2002) reported that PRD1 bacteriophage is more hydrophobic than MS2 bacteriophage. The heterogeneous surface structure of bacteriophage indicates that there could be regions of different degrees of hydrophobicity and charge on the surface.

Table 2.3
Characteristics of bacteriophage used for bench-scale testing

Microbe	Size and shape	Isoelectric point	Zeta potential (mV)*
MS2 bacteriophage	24-26 nm, spherical	3.9	-12.1
PRD1 bacteriophage	65 nm, spherical	4.2	-14.8

Source: Adapted from Herath et al., 1999.
* I = 0.01 at pH 6.5 to 8.5

Table 2.4
Characteristics of microspheres used for bench-scale testing

Microsphere	26 nm	67 nm
Dye	Nile Red	Yellow Green
Isoelectric Point	<2	<2
Zeta Potential (mV)*	-16.1	-17.5
Fluorescence Conditions (Ex/Em)	520/580 nm	490/515 nm
Coefficient of Variation of Diameter	23%	14%
Specific Surface Area (cm^2/g)	2.20×10^6	1
Charge Content Measured from Conductometric Titration (μeq/g)	10.7	5.1
Surface Charge Density ($\mu C/cm^2$)	0.5	0.6
Area Per Charge Group (Å^2/COOH)	3,440	2,760
Charge Groups Per Particle	62	510

* I = 0.01 at pH 6.5 to 8.5

Microsphere Characteristics. Both microspheres were made of polystyrene with carboxyl groups on the surface; they were selected based on size, charge and fluorescent properties. The two microspheres used in this study are described in Table 2.4. They are spherical, and thus, similar in shape to the icosahedrons of PRD1 and MS2 bacteriophage. Additionally, their diameters approximately match the bacteriophage diameters. The charge of the microspheres is caused only by carboxylic acid functional groups, and the charge distribution is similar for both the 26-nm and 67-nm spheres. The zeta potentials for each microsphere were determined at pH 6.5 and pH 8.5. As with the bacteriophage, there was little change over the pH range studied so the two values were averaged in Table 2.4. The 67-nm microspheres were more negative than the 26-nm microspheres over the pH range studied. A major factor differentiating the microspheres from the bacteriophage is that the isoelectric point of the microspheres is reported to be less than pH 2, whereas the isoelectric points of the MS2 and PRD1 bacteriophage were at pH 3.9 and 4.2, respectively.

Table 2.5
Surface characteristics of flat-sheet membranes used for bench-scale testing

	PVDF-hpi	PVDF-hpo	RC-hpi	PES-hpo
Pore Size/MWCO	0.22 μm	0.22 μm	100 kD (0.0073 μm)	100 kD (0.0073 μm)
Contact Angle (°)	18	83	18	58
Zeta Potential (mV)*	-26	-7.0	-4.8	-33
Roughness†	10	10	3	1
Material	PVDF Hydrophilic	PVDF Hydrophobic	Regenerated Cellulose	Polyether-sulfone
Manufacturer	Millipore Durapore	Millipore Durapore	Amicon	Orells
Name	GVWP	GVHP	YM100	PES100

* I = 0.01 at pH 7.0
† Relative ranking based on SEM and AFM

Membrane Characteristics. Four membranes were selected for testing in this study: hydrophilic and hydrophobic PVDF membranes, a hydrophilic regenerated cellulose membrane and a hydrophobic polyethersulfone membrane. Hereafter, these are referred to as PVDF-hpi, PVDF-hpo, RC-hpi, and PES-hpo for identification purposes. For each membrane, the surface was characterized by pore size, contact angle, zeta potential, roughness, and properties resulting from the characteristics of the membrane materials. Additionally, the membranes were hydraulically characterized.

Surface Properties. The membrane surface properties are summarized in Table 2.5. According to the contact angle results, the hydrophobic PVDF membrane is the most hydrophobic, followed by the PES membrane. The two hydrophilic membranes, made from RC and PVDF, have comparable hydrophilicities resulting from equal contact angles. The most negatively charged membranes over the pH range studied are the PVDF-hpi and PES-hpo membranes (Table 2.5). The PVDF-hpo and RC-hpi membranes are also both negative, but only slightly so.

The structures of the membrane polymers are shown in Figure 2.3. The membranes were ranked according to their roughness after visual examination by SEM and AFM (Figures 2.4 and 2.5). The two PVDF membranes were much rougher than the RC and PES membranes, and were thus ranked equally as the most rough. Visual examination of the AFM images showed that the RC-hpi membrane was slightly rougher than the PES-hpo membrane.

Internal Properties. The RC and PES membranes are both asymmetric with a highly porous support layer beneath a tight membrane layer. The PVDF membranes are symmetric, with uniform porosity throughout. The RC and PES membranes would be expected to act as screen filters because their pore size is smaller than the microspheres and bacteriophage, whereas the PVDF membranes are expected to act as depth filters because their pore size is larger than the microspheres and bacteriophage.

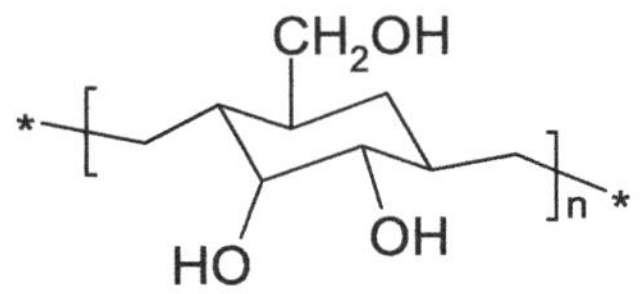

regenerated cellulose (RC)

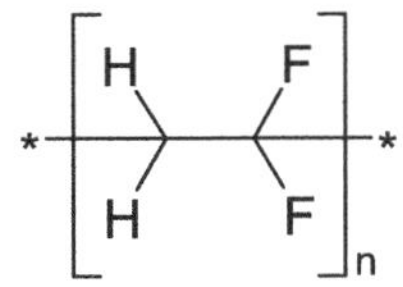

polyvinylidene fluoride (PVDF)

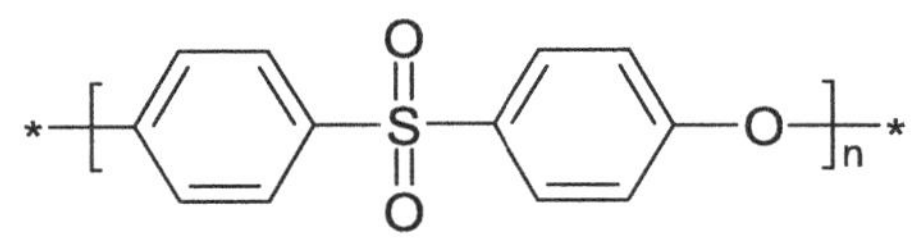

polyethersulfone (PES)

Figure 2.3 Structures of flat-sheet membrane polymer materials used for bench-scale testing

PILOT-SCALE STUDIES

Pilot-Scale Studies in the United States

Site Descriptions and Water Qualities

The main objective of the pilot studies conducted at US facilities was to evaluate low-pressure membranes for the removal of bacteria and viruses under varying water quality and operational conditions. To that end, pilot-scale studies were conducted with two pilot units of different configurations (pressure-driven and submerged membrane systems) by using source waters of varying water qualities. The participating utilities and the source water types used for pilot-scale testing are shown in Table 2.6. As evident from Table 2.6, these source waters covered a wide range of natural water quality parameters that may be encountered in practice.

AFC WTP receives raw water from the Chattahoochee River. The raw water is stored in a 1.4 million m^3 (380 million gallon) reservoir, and then subjected to conventional processes for treatment (coagulation/flocculation/sedimentation process followed by dual media filtration). The raw and filtered waters were used for testing.

Wastewater received at the SWC reclamation facility is treated through a conventional activated sludge process. This product is delivered to various customers for turf irrigation. Excess effluent is treated by microfiltration, followed by reverse osmosis, prior to sub-surface recharge to augment existing groundwater supplies. Only the effluent of the secondary clarifiers (wastewater secondary effluent) was used for testing.

Pilot testing in the US was performed in the following sequence:

- June 2002 to September 2002: testing with the pressurized pilot unit using raw water at the AFC WTP
- October 2002 to January 2003: testing with the pressurized pilot unit using filtered water at the AFC WTP
- February 2003 to April 2003: testing with the submerged pilot unit using raw water at the AFC WTP.

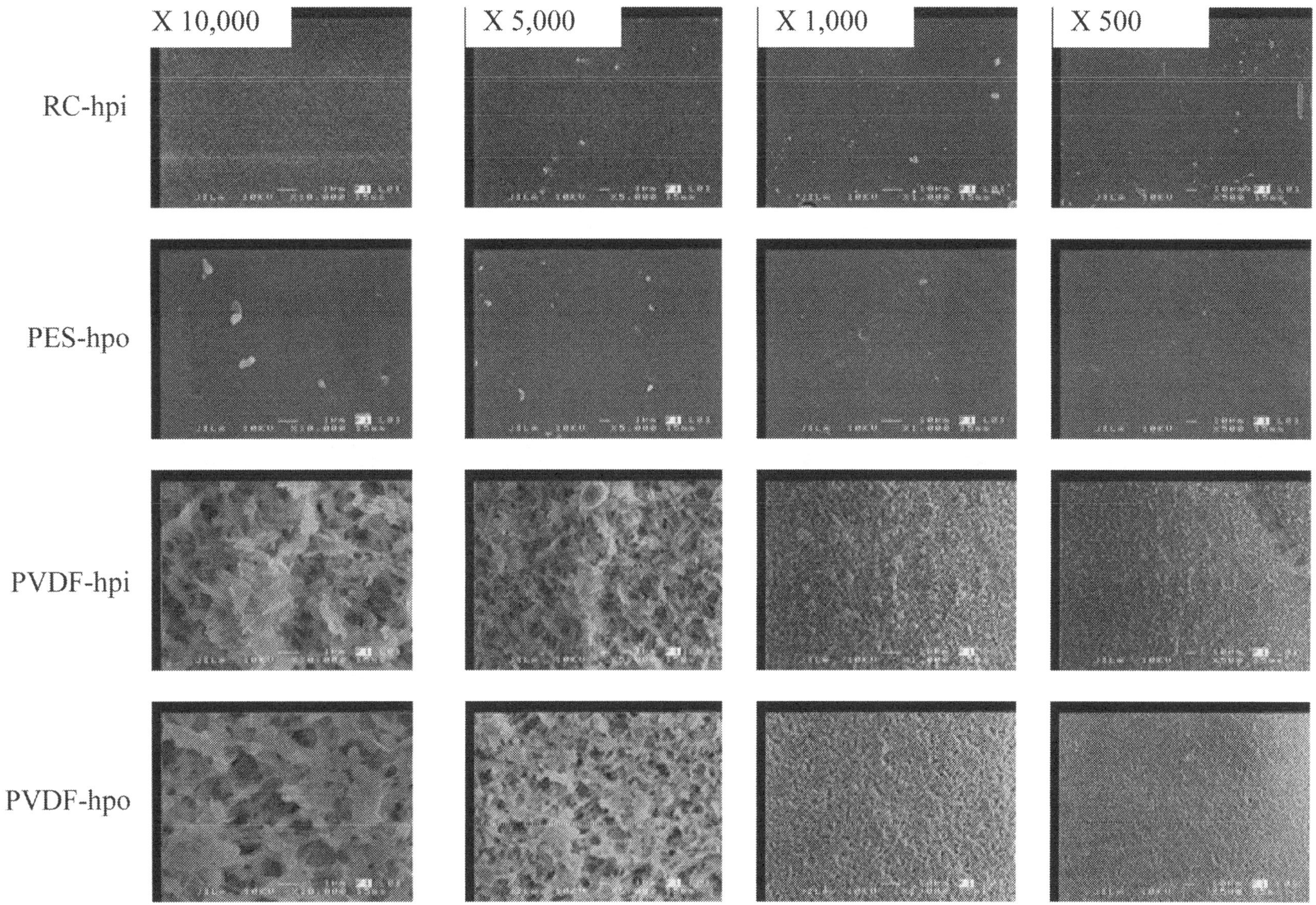

Figure 2.4 SEM images of clean flat-sheet membrane surfaces used for bench-scale testing

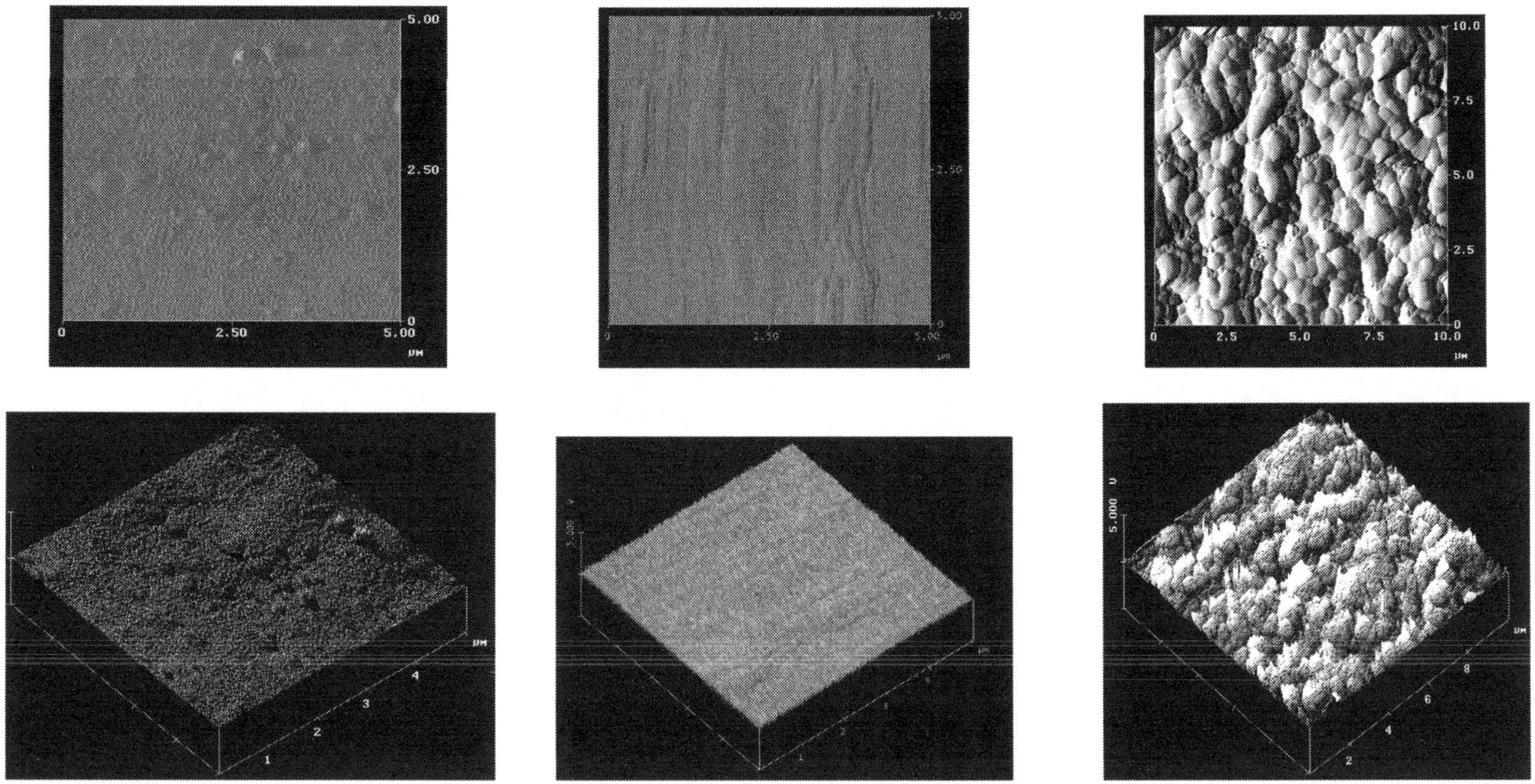

RC-hpi PES-hpo PVDF-hpo

Note: AFM image of PVDF-hpi membrane not available

Figure 2.5 AFM images of flat-sheet membrane surfaces used for bench-scale testing

Table 2.6
Participating utilites and source water qualities for pilot-scale testing

	Atlanta Fulton County Water Treatment Plant		Scottsdale Water Campus
Types of source water	Raw surface water	Filtered surface water	Wastewater secondary effluent
Turbidity (NTU)	2.2-6.8	0.1-0.2	2.1-4.8
Hardness (mg/L $CaCO_3$)	15-18	15-19	298-323
Alkalinity (mg/L $CaCO_3$)	15-18	12-18	174-192
Temperature (°C)	13-19	7-13	32-34
TOC (mg/L)	1.3-1.9	1.1-1.4	5.7-6.4
pH	6.7-7.0	6.2-6.5	7.1-7.3
UV_{254} (cm^{-1})	0.025-0.047	0.015-0.024	106-0.114
SUVA (L/mg-m)	1.5-2.8	1.1-2.2	1.8-2.0

- May 2003 to July 2003: testing with the submerged pilot unit using wastewater secondary effluent at the SWC
- August 2003 to October 2003: testing with the pressurized pilot unit and wastewater secondary effluent at the SWC.

Equipment and Membrane Specifications

Microbial challenges were conducted using the USFilter Memcor 3M10V continuous microfiltration (CMF) pilot unit and the 4S10 submerged continuous (CMF-S) pilot unit. The pressure-driven membrane pilot unit was operated with previously-used PVDF membrane modules. Testing with the submerged membrane pilot unit was initiated with a brand-new membrane at Atlanta. The same membranes were used for testing at AFC WTP and SWC. Table 2.7 shows the characteristics of both membrane pilot units.

The feedwater and filtrate pressures were measured by pressure transducers that were factory-calibrated before initiating the testing at the AFC WTP. The feedwater and filtrate flow rates were measured by digital flow meters that were factory-calibrated before initiating of the testing at the AFC WTP.

The pressurized pilot unit was operated in direct-flow mode, without recirculation of the feed stream. A differential pressure across the membrane allowed filtration to occur. Clean filtrate passed through the fiber wall, along the hollow center of the fiber, and out the filtrate port. The backwash relied on a rapid expansion of gas through the wall of the fiber to remove the contaminants not only from the surface of the fiber, but from the porous membrane matrix as well.

For filtration in the submerged pilot unit, water was drawn through the fiber wall (outside to inside) under suction. All feed passed through the membrane with no crossflow used while in filtration mode. The backwash used air to scour the fibers, while a small amount of filtrate was pushed backward through the fibers (inside to outside) to remove the layer of accumulated

Table 2.7
Characteristics of the pressurized and submerged membrane pilot units used in US testing

Characteristics	Pressurized pilot unit	Submerged pilot unit
Number of modules	3	4
Operation	Pressure driven, dead end filtration	Submerged, vacuum driven, dead end filtration
Feed flow configuration	Outside → Inside	Outside → Inside
Membrane material	PVDF	PVDF
Membrane geometry	Hollow fibers	Hollow fibers
Filtration surface per module	28.7 m^2 (309 ft^2)	31.1 m^2 (335 ft^2)
Nominal pore size	0.1 μm	0.1 μm
Potting Material	Polyurethane	Polyurethane
Transmembrane Pressure (P_{tm})*	0.28 to 1.52 bar (4 to 22 psi)	0.28 to 0.97 bar (4 to 14 psi)
Clean Water Flux (J) †	18 L/m^2/hr to 54 L/m^2/hr (11 gfd to 32 gfd)	14 L/m^2/hr to 50 L/m^2/hr (8 gfd to 29 gfd)
Backwashing duration/frequency	2.5 minutes/hourly	2.5 minutes/hourly

* $P_{tm} = P_i - P_o$ (bar or psi)
P_i: pressure at the inlet of the membrane module (bar or psi)
P_o: pressure at the outlet of the membrane module (bar or psi)

† $J = \frac{Q}{A}$ (L/hr/m^2 or gal/ft^2/day) where:
Q: filtrate flow (L/hr or gal/day)
A: membrane surface area (m^2 or ft^2)

material The dirty backwash water was drained away, and the tank was refilled prior to recommencing filtration.

The filtration and the backwash cycles in both membrane pilot units were automated by use of a programmable logic controller (PLC). The program stored within the PLC memory controlled the solenoid valves, the panel indicators and the feed pump. A second regulator mounted on the pneumatic panel regulated air pressure for backwash and rewet cycles. A third regulator provided the proper pressure for operating the membrane test function.

Figures 2.6 and 2.7 respectively show the flow diagrams of the pressurized membrane pilot unit and submerged membrane pilot unit in filtration mode (backwash system not shown). Figure 2.8 is a photograph of both units at the AFC WTP.

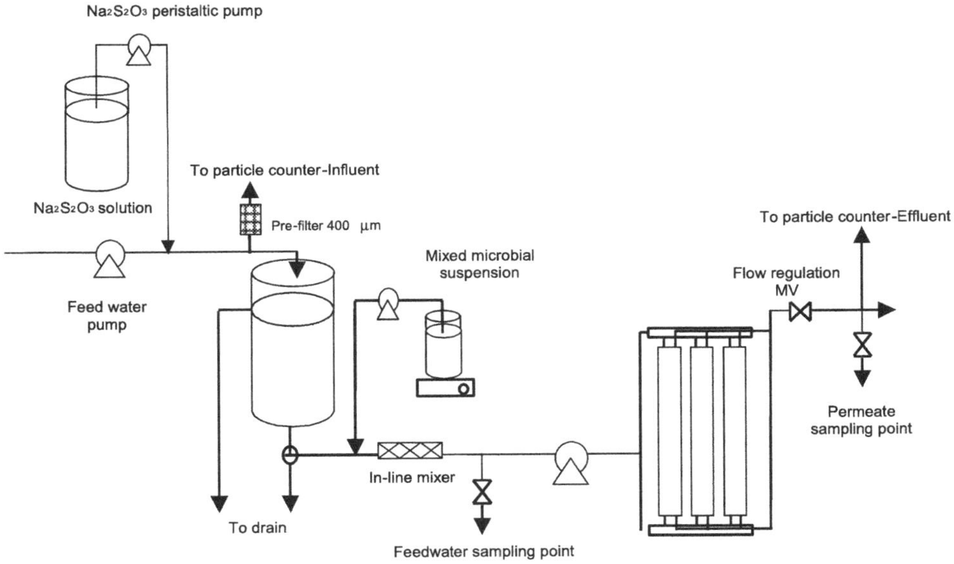

Figure 2.6 Flow diagram of the pressurized membrane pilot unit in filtration mode (pilot testing, US)

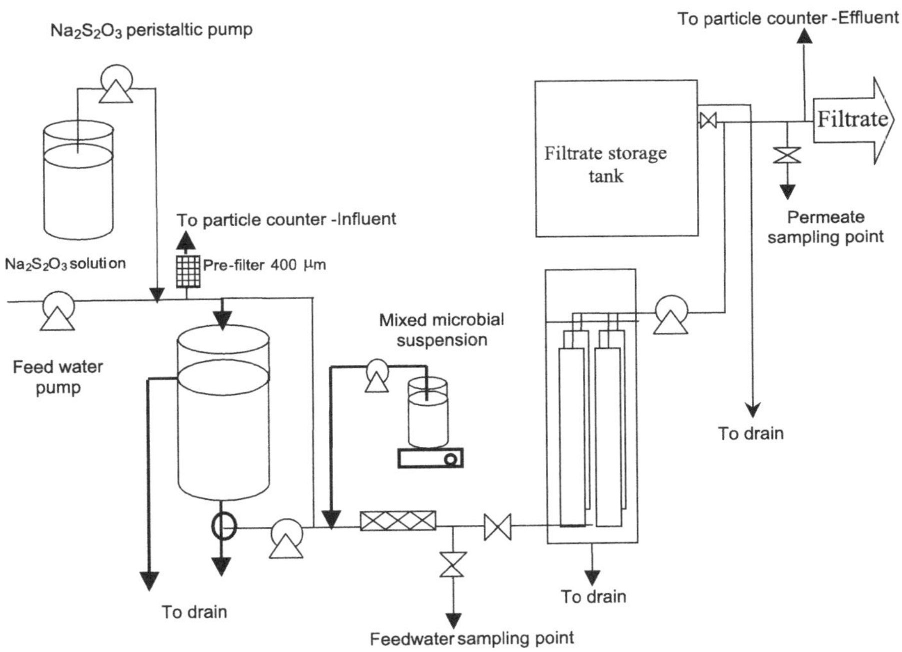

Figure 2.7 Flow diagram of the submerged membrane pilot unit in filtration mode (pilot testing, US)

Figure 2.8 Pressurized (left) and submerged (right) membrane pilot units at the AFC WTP (pilot testing, US)

Description of Studies

An experimental protocol was developed for conducting microbial challenge testing at pilot scale. The protocol was comprised of 4 tasks:

- Task 1: Hydraulic characterization
- Task 2: Microbial removal efficiency under clean water conditions
- Task 3: Evaluation of the impact of deposition layer formation on microbial removal
- Task 4: Evaluation of the impact of membrane fouling on microbial removal

The hydraulic characterization of both units was performed at the AFC WTP. Microbial removal efficiency under clean water conditions, evaluation of the impact of deposition layer formation on microbial removal and evaluation of the impact of membrane fouling on microbial removal were conducted with raw and filtered water at the AFC WTP as well as with secondary effluent at the SWC.

Task 1: Hydraulic Characterization. Task 1 consisted of two elements: membrane integrity testing to verify membrane integrity and tracer testing to characterize the residence time distribution.

Integrity Testing. The integrity of the pressurized and submerged systems was checked through a pressure decay test (PDT) after each set of microbial challenges. A PDT consisted of draining the filter lumens and pressurizing the filtrate side of the membranes to approximately 1 bar (15 psi) for about 10 seconds. The unit was stabilized (pressure line closed) for 120 seconds to allow for pressure decay testing. A pressure decay greater than 0.014 bar/min (0.2 psi/min) was indicative of a compromised module (e.g., hole(s) in fibers or potting failure). When the pressure decay test exceeded 0.014 bar/min (0.2 psi/min), each module was inspected using a pin repair kit. Once identified, the compromised fiber/fibers was/were plugged either at both ends for the pressurized pilot unit or at the single end for the submerged pilot unit.

Tracer Testing. Hydraulic characterization consisted of conducting tracer studies to determine the residence time distribution of both the pressurized and submerged pilot units. Tracer tests were performed to determine the number of hydraulic retention times (HRTs) to reach steady-state conditions for each membrane pilot unit. Before conducting each tracer test, each membrane pilot unit was chemically cleaned according to the manufacturer's instructions. After the chemical cleaning, a short rinse (10 to 15 min) with raw water was performed, with feedwater and filtrate samples then analyzed for conductivity to verify that chemical cleaning did not affect conductivity. Tracer tests were performed for each membrane pilot unit in a dead-end filtration mode at each flow rate at which mircorobial challenges were conducted. The sodium chloride dose was 250 mg/L. Sodium chloride granules were dissolved in warm water and mixed overnight to ensure complete dissolution. The microbial feed pump was used to inject the sodium chloride solution upstream of the in-line mixer (see Figures 2.6 and 2.7). Upon initiating the step addition of sodium chloride solution at a constant dose, samples were collected at the feedwater and permeate sampling points (see Figures 2.6 and 2.7) for conductivity analysis. Four influent samples were collected during each tracer test, while permeate samples were collected every 15 s. Samples were collected for 5 to 10 min, depending on the retention volume and flow rate of the membrane pilot unit. Samples were collected into labeled containers then analyzed with a portable conductivity meter. T_{99} represents the time required for 99% of the incoming tracer concentration to exit the contactor using a pulse-input. The number of HRTs to reach steady-state conditions was calculated as follows:

$$\text{number of HRTs} = \frac{T_{99}}{HRT} \tag{2.4}$$

with: $$HRT = \frac{\text{Liquid volume of the membrane unit}}{\text{Flowrate during tracer test}}$$

Task 2: Microbial Removal Efficiency Under Clean Water Conditions. The microbial challenges to quantify removal efficiencies were conducted with dechlorinated tap water at both the AFC WTP and the SWC. The objective of these challenges was to quantify the removal efficiency of each membrane system under the worst-case conditions from a microbial removal perspective (i.e., no deposition layer accumulation and no fouling).

Before initiating each seeding challenge, a filtrate sample was collected to ascertain the absence of microbial contamination. Free chlorine residual in the finished water at the AFC WTP was quenched with sodium thiosulfate ($Na_2S_2O_3$) prior to entering the holding tank of the pilot unit. Conductivity measurements indicated that there was no significant impact of the addition of sodium thiosulfate. The free chlorine residual was carefully monitored in the holding tank before starting the seeding of microorganisms to ensure that the water was free of chlorine. Once a chlorine-free environment was ensured, a mixed suspension of MS2 bacteriophage, PRD1 bacteriophage, and *Pseudomonas diminuta* bacteria was seeded into the system at the desired flow rate through a peristaltic pump. Microbial challenges were performed at three different flow rates within the same filtration cycle (i.e., no backwash occurred during the whole test). The filtration flux was initiated at 11 L/m^2/hr (93 gfd), then increased to 27 L/m^2/hr (228 gfd), and finally to 37 L/m^2/hr (312 gfd). Feedwater and filtrate samples were collected at each flow rate. Challenges were repeated under identical conditions.

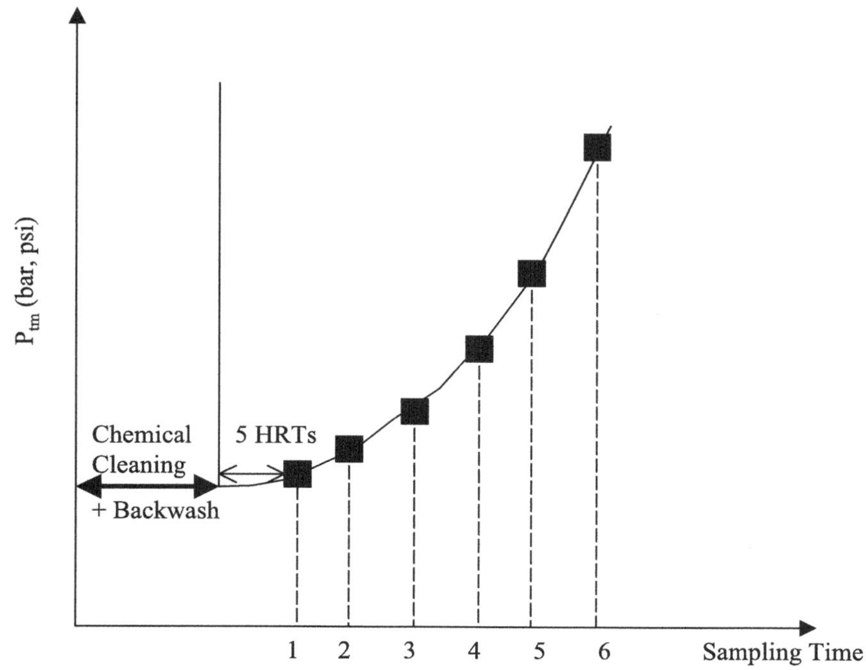

Figure 2.9 Sequence of testing for evaluation of the impact of deposition layer formation on microbial removal

Task 3: Evaluation of the Impact of Deposition Layer Formation on Microbial Removal. The evaluation of the impact of deposition layer formation on microbial removal consisted of short-term experiments during which the membrane was operated over 4 hours without back-washing, allowing any deposition layer accumulate and quantifying the microbial removal at various time intervals over the filtration cycle.

For each source water, three microbial challenges were conducted at different fluxes (11, 27 and 37 L/m^2/hr; 93, 228 and 312 gfd) using MS2 and PRD1 bacteriophage seeded into the test water. The experiments were started immediately after chemical cleaning and backwashing of the membranes. Therefore, the deposition layer and fouling on the membrane surface was reduced to its minimal level at the beginning of each experiment. The first filtrate sample for microbial enumeration was collected as soon as steady-state conditions were reached, and five subsequent samples were collected at regular intervals over the four-hour filtration cycle. Figure 2.9 depicts the sequence of the microbial challenges for this type of experiment. A minimum of two feed-water samples was also collected to verify the seeded concentration of microorganisms, usually at the beginning and end of each challenge experiment.

Task 4: Evaluation of the Impact of Membrane Fouling on Microbial Removal. Evaluation of the impact of membrane fouling on microbial removal consisted of long-term experiments, allowing the membranes to operate over 10 to 15 days with regular backwashing, and quantifying the microbial removal at various times over the that period.

Initiation of an experiment started immediately after chemical cleaning and backwashing of the membranes. Flow rates and backwash intervals were determined according to the anticipated increase in transmembrane pressure and the impact of irreversible fouling on the transmembrane

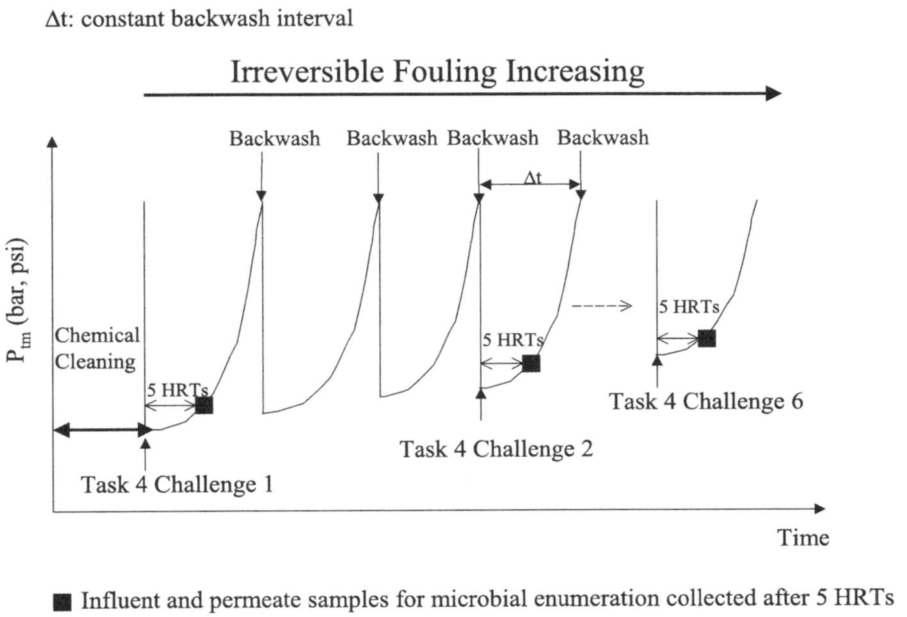

Figure 2.10 Sequence of testing for evaluation of the impact of membrane fouling on microbial removal

pressure. The unit was allowed to operate continuously without changing the flow rate or the backwash interval. Six seeding challenges, using MS2 and PRD1 bacteriophage, were performed over the duration of the experiment as the transmembrane pressure increased. Microbial challenges were initiated immediately after backwashing to ensure that deposition layer formation was not a contributing factor to removal of the microorganisms. For each microbial challenge, one feedwater and one filtrate sample were collected and water quality parameters were recorded. Figure 2.10 depicts the sequence of the microbial challenges for this type of experiment.

Calculation of Microbial Removal

Operating a clean membrane with clean water is the most favorable condition for passage of bacteriophage as compared to a fouled system, where mechanisms such as deposition layer formation and/or fouling lead to pore restriction, thus potentially enhancing the microbial removal. The microbial removal recorded under clean water conditions (Task 2) served as the baseline for calculating microbial removal for each membrane under conditions of deposition layer formation and fouling. The microbial removal occurring during the subsequent tasks (Task 3, deposition layer formation; and Task 4, fouling) were reported as net microbial removal using the following equation:

$$\text{net microbial removal} = (\text{Microbial removal})_{FW} - (\text{Microbial removal})_{CW} \quad (2.5)$$

where:
$(\text{Microbial removal})_{FW}$: Microbial removal observed with the feed water (raw or filtered water)
$(\text{Microbial removal})_{CW}$: Microbial removal value observed under clean water conditions

This calculation allowed differentiation of the additional removal induced by deposition layer formation or membrane fouling from the intrinsic removal capacity of the membrane.

Water Quality Analysis and Operational Parameters

For each microbial challenge, the following feedwater quality parameters were monitored: pH, alkalinity, hardness, turbidity, temperature, TOC, UV absorbance at 254 nm, total suspended solids and particle counts. All analyses were performed according to Standard Methods (APHA, et al. 1998). Operational parameters such as filtrate flow rate and transmembrane pressure were also recorded during each microbial challenge.

To compare the hydraulic performance of each membrane system from one site to another, the specific flux was temperature-corrected to 20°C according to the following equation (Lide, 2004):

$$J_{S,20} = J_{S,T} \times \exp[-0.029 \times (T - 20)] \qquad (2.6)$$

where: $J_{S,20}$ is the specific flux (L/m^2/hr/bar) at 20°C
$J_{S,T}$ is the specific flux measured at temperature T
T is the actual temperature (°C)

Cleaning and Disinfection Procedures

Prior to conducting a set of microbial challenges, chemical cleaning of the units was performed in order to restore membrane productivity/efficiency. The chemical cleaning for both the pressurized and submerged units consisted of two successive cycles:

1. acid cleaning with citric acid (2% solution by weight)
2. chlorine cleaning with sodium hypochlorite (400 mg/L chlorine solution).

For the pressurized unit, the sequence of each cleaning cycle was:

1. Filtrate recirculation for 30 minutes: The solution was filtered through the membranes and recirculated.
2. Soaking for 30 minutes: The membranes were allowed to soak in the solution.
3. Shell recirculation for 60 minutes: The solution is recirculated on the shell side of the membranes, with the filtrate side closed.

No chemical cleaning or disinfection was performed between each microbial challenge of the evaluation of the impact of membrane fouling on microbial removal (Task 4), since the chemical cleaning and disinfection procedures could have impacted the fouling of the membranes.

In addition to the chemical cleaning described above, the units were also disinfected prior to each set of microbial challenges. Chlorinated water was filtered for an hour and the chlorine residual (at least 2 mg/L) was allowed to stand in the unit overnight. The unit was then flushed with water containing sodium thiosulfate to quench any remaining chlorine residual. The submerged unit appeared to be more susceptible to biofilm growth than the pressurized unit due to

Table 2.8
Size of microorganisms used for seeding challenges

Type of microorganism	Microorganism	Approximate size (μm)	Reference
Bacteria	*Pseudomonas diminuta*	0.3-0.4 * 0.6-1	ASTM F838-83, 1993
Bacteria	*Klebsiella terrigena*	1-1.5 * 0.5	Izard et al., 1981
Virus	PRD1 bacteriophage	0.07	Hunter, 1997
Virus	MS2 bacteriophage	0.027	Joklik et al., 1992

a more complex piping arrangement and a higher potential for dead zones, which made disinfection more difficult. Therefore, a higher chlorine dose was applied to the submerged unit in an enhanced disinfection procedure. Chlorinated water at a concentration of 20 mg/L was filtered for an hour and allowed to stand in the unit overnight (after which the chlorine residual in the effluent was at least 15 mg/L).

Microbial Challenge Methodology

Microbial challenges were conducted with MS2 and PRD1 bacteriophage as viral surrogates and *Pseudomonas diminuta* and *Klebsiella terrigena* as bacterial surrogates. Table 2.8 presents the characteristics of these microorganisms.

Microbial stock suspensions were prepared at the USF and shipped overnight to the site where they were stored in a refrigerator at 4°C. *Pseudomonas diminuta* and *Klebsiella terrigena* stock suspensions were prepared and shipped a few days before the actual microbial challenge. MS2 and PRD1 bacteriophage were stored on site for a maximum duration of 3 months.

Prior to each set of microbial challenges, a pressure decay test was performed as described above. Microbial challenges were conducted using a continuous seeding system with a well-mixed suspension of microorganisms injected through a peristaltic pump into the feedwater line which contained a static mixer (see Figures 2.6 and 2.7). All organisms were seeded as a cocktail in order to evaluate their removal under the same water quality and operating conditions. The microbial suspension was prepared by dilution of microbe stocks with the feedwater (raw or filtered water at the AFC WTP, secondary effluent at the SWC). The flow rate of the microbial suspension was adjusted in accordance with the feedwater flow rate to keep the microbial concentration constant. A minimum concentration of 10^5/mL of each organism was maintained in the seeded feedwater. The flow rate of the microbial suspension ranged from 0.1 to 0.2 L/min. For the microbial challenges conducted with the pressurized and submerged units, 6 L of microbial suspension were prepared. An exception was the deposition layer formation challenges for the submerged unit, for which 12 L of microbial suspension were prepared due to the longer operation time to reach steady-state conditions.

For each challenge, feedwater and filtrate samples were collected after steady-state conditions were reached, as established by the tracer studies. Sampling ports were intensely flushed and propane flamed prior to each sampling. Each sample was collected in a sterile 50-mL polypropylene centrifuge tube. Control samples consisting of a diluted sample of the microbial stock

Figure 2.11 Submerged pilot unit at the KPAWTW

suspensions were included with all samples that were shipped to USF for enumeration. The objective of this sample was to quantify any changes in the microbial concentration due to shipping and handling conditions not related to experimentation. Microbial samples were stored in a cooler with blue-ice packs upon collection and shipped overnight to USF. Microbial samples were prepared at three different dilution ratios, and each dilution was plated and enumerated in duplicate as described below.

Pilot-Scale Studies in the United Kingdom

Site Descriptions and Water Qualities

The UK pilot studies were performed to evaluate the ability of two membrane types to remove bacteria and bacteriophage (surrogate for viruses) from different water sources and under different operational conditions. Figures 2.11, 2.12, and 2.13 show pilot units for the UK testing. These two membrane types comprised a submerged system and a pressurized system. The source waters used for the UK testing were raw water and filtered water. The raw water was supplied by the Queen Mary Reservoir fed from the river Thames. The filtered water was supplied by Kempton Park Advanced Water Treatment Works (KPAWTW) and was treated by slow sand filtration and chloramine disinfection, then dechlorinated before use in the pilot plant. Table 2.9 summarizes the range of water quality parameters for each water type.

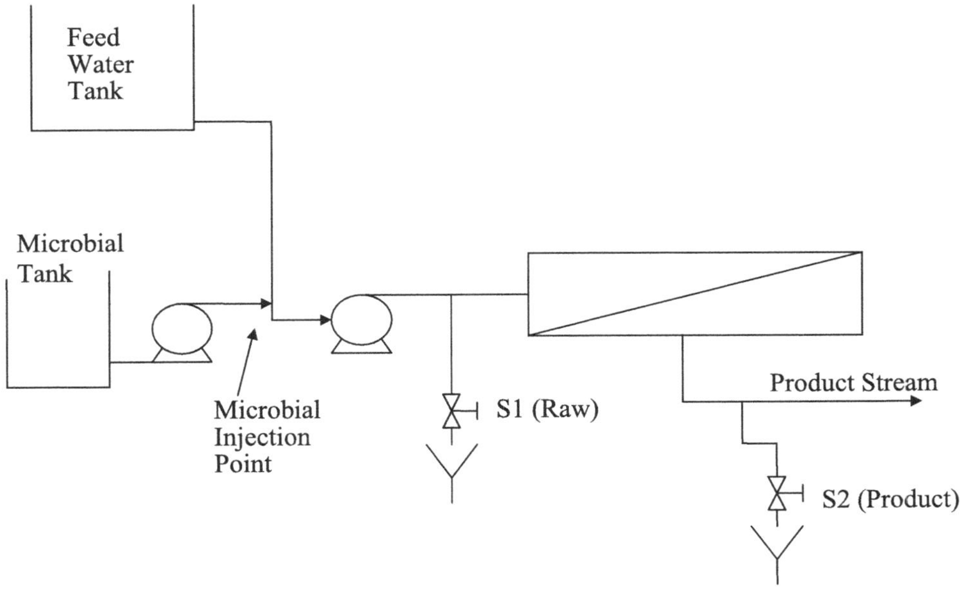

Figure 2.12 Schematic diagram of the pressurized pilot unit (pilot testing, UK)

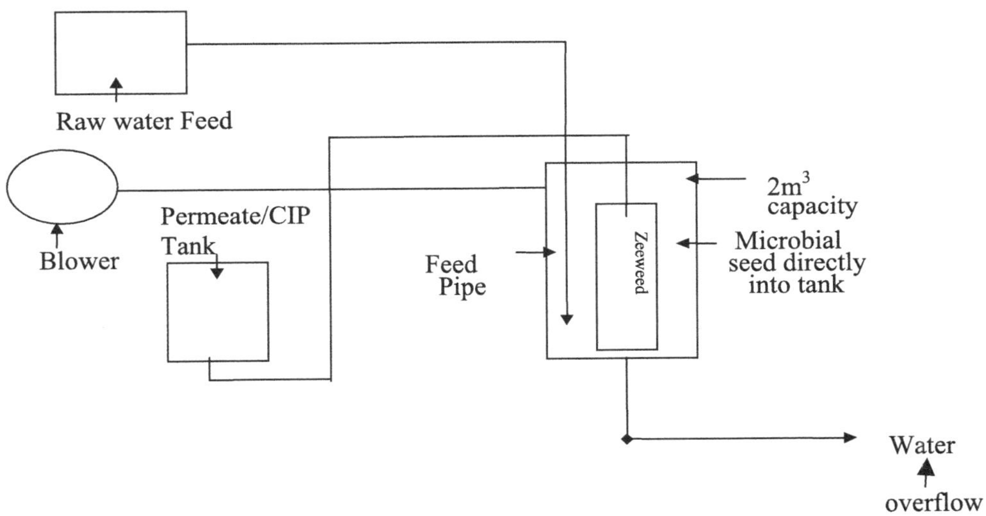

Figure 2.13 Schematic diagram of the submerged pilot unit (pilot testing, UK)

Pilot testing at KPAWTW was performed in the following sequence:

- April 2003 to July 2003: Testing with the submerged pilot unit using raw water
- April 2003 to July 2003: Testing with the submerged pilot unit using finished water
- August 2003 to November 2003: Testing with the pressurized pilot unit using finished water
- November 2003: Testing with the pressurized pilot unit using raw water

Table 2.9
Source water quality for UK pilot testing

Parameter	Raw water	Filtered water
Turbidity (NTU)	1.7-3.2	<0.1-0.18
Hardness (mg/L$CaCO_3$)	265-297	266-329
Alkalinity (mg/L $CaCO_3$)	202-224	134-223
Temperature (°C)	11-21	11-18
pH	8.2-8.5	7.1-7.9
UV_{254} (cm^{-1})	0.07-0.08	0.02-0.05
TOC (mg/L)	4.4-5.1	2.7-3.4

Equipment and Membrane Specifications

Characteristics of the Submerged Pilot Unit. The pilot plant incorporated a single membrane cassette submerged in a filtration tank. See Table 2.10. Feedwater was added continuously into the base of the tank (even during a backwash), creating an upward flow through the tank with excess or "reject" flow spilling out through an overflow. Filtrate was drawn through the membrane by a negative (vacuum) pressure created on the product side of the membranes by a vacuum pump. The membrane used was a hollow fiber with the flow direction from the outside to the inside.

Material removed by the membrane was retained on the feed side of the membrane within the filtration tank. Backwash was achieved using product water to back flush the membrane, while simultaneously bubbling air upward from the base of the tank such that bubbles were directed between and around the membrane fibers. The air bubbles had the dual function of agitating and lifting the membrane fibers. The air agitation combined with the back flush encouraged the physical removal of accumulated material from the membrane surface. In addition, the air stirred up sediment within the tank, and the upward flow assisted in moving sediment out through the overflow during backwash.

Chemical cleaning was performed in the pilot plant either by a chemically enhanced (chlorine) backwash (via PLC control) or by taking the process off line and performing a maintenance cleaning (also under PLC control) using either an acidic or a caustic solution. The chemical was backflushed through the membrane and left to stand for 60 minutes. The tank was drained and then rinsed several times by backwashing feed water through the membrane until all traces of chemical were removed.

The plant production was controlled to a constant pre-set product water output, requiring an increasing transmembrane pressure (P_{tm}) as the membrane became fouled. P_{tm} was allowed to increase to a maximum of 0.4 bar (5.6 psi) negative pressure. The plant automatically shut down when the maximum P_{tm} was reached, and a chemical cleaning was performed.

The pilot plant was fully automatic, with operator access to change set points and timers governing product water flow rate, backwash flow rate, backwash interval and duration, and chemical cleaning interval.

Table 2.10
Characteristics of the submerged membrane used in UK pilot testing

Membrane specifications	
Module type	Zeeweed 500b
Configuration	Outside/In Hollow Fiber
Number of modules	One
Membrane type and total surface area of module	Polypropylene, 46.5m^3
Nominal pore diameter	0.035 µm
Volume of membrane module	5.9 liters
Operating limits	
Maximum P_{tm}	0.83 bar @ 40°C
Typical operating P_{tm}	0.07-0.55 bar @ 40°C
Maximum operating temperature	40°C
Operating pH range	pH 5-9
Maximum cleaning temperature	40°C
Cleaning pH range	pH 2-11 (max. 1 hr/day)
Maximum chlorine tolerance	1,000 mg/L (max. 1 hr/day)
Maximum transmembrane backwash pressure	0.55 bar
Aeration flow	8.5-26 m^3/h
Flow rate	2.2-4.0 m^3/h
Clean water flux	47-86 L/m^2/h
Backwashing duration/frequency	30 sec/15 min

Characteristics of the Pressurized Pilot Unit. The pressurized pilot unit was also a hollow fiber system, containing two membranes in series. See Table 2.11. Water entered the inside of the fiber lumen and was forced out through the membrane (i.e., inside to outside filtration). The pilot unit was supplied with dechlorinated tap water (slow-sand filtered water from the water treatment plant that had been chloraminated and then dechlorinated) or raw reservoir water that had undergone pre-filtration through rapid gravity filter columns (with no upstream chemical dosing, no flocculation, and no pre-filtration chemical addition). The flow rate through the rapid gravity filter columns was 11.8 m^3/h (52 gpm). The filter columns were 440 mm (17.3 in) in diameter and 600 mm (24.6 in) deep. The media consisted of sand, pumice and anthracite with media sizes between 0.6 and 1.2 mm. The rapid gravity filters were automatically backwashed. The pre-treatment stage was necessary to remove large particulates that could potentially plug the lumen of the hollow fibers. The pilot plant was fed from a balance tank with a volume of approximately 6 m^3 (1,585 gal). The balance tank provided equalization of the supply to the membrane unit during rapid gravity filter backwash. The feedwater entered the pilot plant through a 0.1-mm strainer to supply flooded suction to the feed pump that provided the operational pressure and flow rate required. The inlet turbidity and flow rate were monitored; the flow

Table 2.11
Characteristics of the pressurized membrane used in UK pilot testing

Membrane specifications	
Module type	X-Flow Xiga membranes
Membrane chemistry	Polyethersulphone (PES)/ polyvinyl pyrrolidone (PVP)
Fiber internal diameter	0.8 mm
Pore size	0.025-0.03 μm
Molecular weight cut-off	150 kD
Element length	1.5 m
Membrane area	35 m^2
Number of fibers per element	10,000

rate was set by manual initiation via the PLC. The inlet flow line divided after the pump to feed the membrane pressure vessel from both ends. The water then passed through the membrane and was collected through manifold connections also at both ends of the vessel. The filtrate was directed to a backwash tank, with tank level maintained by waste overflow.

The inlet pressure was monitored at both inlet ends, and the membrane outlet pressure was monitored at both outlet ends of the pressure vessel. Filtrate turbidity was also monitored. Calibration of all instruments was performed on a monthly basis.

The plant could be run in a fully manual mode or automatically under control of the PLC. The backwash interval was set manually. In all experiments, backwash interval was set to 15 minutes with a duration of 30 seconds.

Filtered water used as feedwater for the pilot plant was dechlorinated by dosing sodium bisulfite at the minimum concentration necessary to remove the chlorine residual otherwise found in the supply. The dechlorination chemical was applied downstream of the feed pump. The supply to the membrane pilot plant was checked manually prior to each microbial challenge to ensure a chlorine-free feedwater.

Description of Studies

The pilot studies performed in the UK comprised the same four tasks performed in US pilot testing. The exceptions to the experimental protocol described above for US testing are summarized below.

Task 1: Hydraulic Characterization. Integrity testing was performed on both submerged and pressurized pilot units by conducting PDTs. A PDT consisted of draining the filter lumens and pressurizing the filtrate side to approximately 0.3 bar (4.4 psi) as recommended by the manufacturer. The pressure decay was recorded at 10-second intervals. A PDT was performed at the beginning of each experiment. No indication of compromised membranes was observed.

For the submerged membrane unit, tracer tests were conducted at three fluxes, 49, 73 and 86 L/m^2/hr (414, 616 and 726 gfd). Tracer tests were conducted in the pressurized pilot unit at fluxes of 43, 70 and 96 L/m^2/hr (363, 591 and 810 gfd).

Task 2: Microbial Removal Efficiency Under Clean Water Conditions. For the submerged pilot unit, each set of microbial challenges was performed in duplicate using dechlorinated tap water at three different fluxes, starting at 49 L/m^2/h (414 gfd), then increasing to 73 L/m^2/h (616 gfd) and finally to 86 L/m^2/h (726 gfd) with no backwash between each change in flux. For the pressurized pilot unit, each set of microbial challenges was performed in duplicate at fluxes of 43, 71 and 96 L/m^2/hr (363, 591 and 810 gfd) with no backwash between each change in flux.

Task 3: Evaluation of the Impact of Cake Layer Formation on Microbial Removal. For the submerged pilot unit, the experiments were performed using raw water or filtered water over a 4-hour period at fluxes of 49, 73 and to 86 L/m^2/h (414, 616 and 726 gfd). Microbial challenges using MS2 and PRD1 bacteriophage only were performed at six time intervals over the filtration cycle (every 44 min) at each flux. For the pressurized pilot unit, experiments using raw or filtered water were performed over a 4-hour period at fluxes of 43, 71 and 96 L/m^2/hr (363, 591 and 810 gfd).

Task 4: Evaluation of the Impact of Membrane Fouling on Microbial. For the submerged pilot unit, two experiments were performed with raw water. The unit was operated until the maximum pressure was reached and automatic shutdown occurred. Both experiments were performed over four days with four microbial challenges during the 4-day period. An additional experiment was performed with filtered water over a 7-day period, with six microbial challenges over that time. All microbial challenges were performed using MS2 and PRD1 bacteriophage.

For the pressurized pilot unit, both the raw water and filtered water experiments were performed over five days, with five microbial challenges performed over each 5-day experiment.

Cleaning and Decontamination Procedures

For the submerged pilot unit, sufficient sodium hypochlorite was added to the filtrate/CIP tank to provide a 200-mg/L free chlorine residual. This solution was backwashed through the membranes into the filtration tank to ensure thorough cleaning and disinfection of the membrane, the filtrate/CIP tank and the product water lines. The chlorine concentration in the filtration tank was then increased to provide a free chlorine residual of approximately 100 mg/L, which was allowed to stand for approximately 1 hour before the pilot unit was drained and rinsed with feed water until all traces of chlorine were removed. The absence of chlorine was confirmed by analysis for chlorine residual before the plant was brought back into service.

For the pressurized pilot unit, two chemical cleaning solutions were used. Before and after each experiment involving microbial challenges, the membranes were cleaned with sodium hypochlorite at a 100-mg/L free chlorine residual. In addition, due to the hardness of the raw and filtered water, an acid clean (2% citric acid, adjusted to pH 3 using sulfuric acid) was also required as general maintenance to help prevent scale formation. The appropriate chemical was added to the backwash tank and the membrane was backwashed under manual control. The plant was then shut down while still full of chemical solution and allowed to stand for 1 hour. The membranes were then flushed free of chemical and brought back into service.

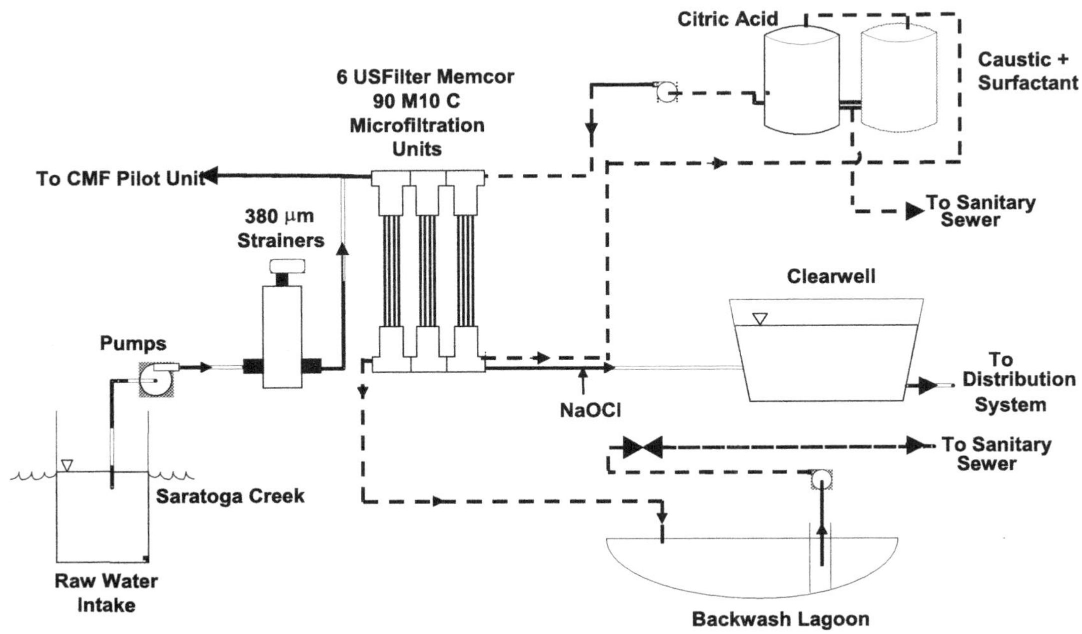

Figure 2.14 Flow diagram of the full-scale Saratoga (CA) Filtration Plant

FULL-SCALE STUDY

Plant Description and Water Quality

Full-scale testing was conducted with raw surface water from the Saratoga Creek River in Saratoga, CA. The Saratoga Filtration Plant (SFP) receives raw water from the Saratoga Creek and subjects it to direct membrane filtration through a pressure-driven membrane system (USFilter Memcor CMF) and chlorine disinfection (see Figure 2.14). The nominal flow rate of the Saratoga Filtration Plant is 790 m^3/hr (5 mgd). The full-scale membrane unit is composed of 6 membrane racks, each comprising 90 polypropylene membrane modules. The pressurized pilot unit was installed and operated side-by-side with the full-scale membrane process.

Description of Study

The objective of the full-scale study conducted at the Saratoga Filtration Plant was to quantify the microbial removal of identical polypropylene modules operated similarly on a full-scale pressure-driven membrane system and the pilot unit. The specific goals of the full-scale testing were to:

- monitor and compare fouling rates of membrane modules of same age (brand-new at the beginning of the period of testing) operated on a pilot-scale unit and a full-scale system
- quantify the microbial removal of these two modules over the period of testing
- compare the microbial removal of the brand-new modules and mature modules.

Three modules were used during the full-scale testing: one mature module that had been in use on the full-scale plant for five years, and two brand-new polypropylene modules. One brand-new module was installed in the pilot unit (designated "pilot-scale new module" or "PSN") and the two other modules were used in the full-scale system (designated "full-scale new module" or "FSN," and "full-scale mature module" or "FSM").

The pilot- and full-scale units were operated concurrently for nine weeks under the same operating conditions. The module in the pilot unit was continuously operated at a similar flux to those modules used in the full-scale plant; and backwashing and chemical cleaning were performed on both units at the same frequency.

Every two weeks, the FSN and FSM modules were removed from the full-scale system and installed in the pilot unit in order to conduct a microbial challenge. Modules used in the pilot unit could be isolated from filtration, which allowed challenging the modules individually. Four microbial challenges were conducted over the nine-week period at a flux of 34 $L/m^2/hr$ (20 gfd) using MS2 and PRD1 bacteriophage seeded into the raw water. Microbial challenges were conducted on the pilot unit only, and only one module was operated at a time. Prior to each set of microbial challenges, a pressure decay test was performed on the pilot unit operated with the three test modules (PSN, FSN, and FSM) to ascertain their integrity.

After being challenged with microorganisms, the modules were replaced on the full-scale system. For each challenge, feedwater and filtrate samples were collected, and water quality parameters were recorded (turbidity, TOC, particle counts, and UV_{254}). Modules were operated for two weeks between every microbial seeding with backwashes every 45 minutes, which allowed the bacteriophage retained during the microbial challenges in the pilot plant to be removed before the next challenge. Background samples collected in the pilot plant filtrate just prior to the actual seeding confirmed the absence of previously seeded bacteriophage.

Prior to initiating the nine-week experiment, two microbial challenges were conducted in the pilot unit using microfiltered water from the full-scale plant at a flux of 34 $L/m^2/hr$ (20 gfd). These two challenges aimed at quantifying the initial removal performance of each membrane module under the worst conditions from a microbial removal perspective (no additional fouling during the microbial challenge). A suspension of MS2 and PRD1 bacteriophage and *P. diminuta* was spiked into the system at the desired flow rate. Feedwater and filtrate samples were collected.

MICROBIAL PROPAGATION AND ANALYSIS

Bacteriophage

Bacteriophage stocks were shipped via overnight delivery service to each pilot facility at the time of testing. Stocks were maintained at approximately 4°C until used and were never held for longer than 4 months. Trip control samples consisting of the refrigerated bacteriophage stocks were shipped with each set of experimental samples to assure that there was no reduction in bacteriophage concentration during storage or shipping.

MS2 and PRD1 bacteriophage were enumerated according to National Water Research Institute and Awwa Research Foundation (NWRI and AwwaRF 2000). MS2 bacteriophage was received from the American Type Culture Collection (ATCC, Manassas, Virginia) ATCC #15597-B1. The *Escherichia coli* C-3000 – ATCC 15597 was utilized as the host bacterium, with the bacterial growth media being tryptic soy broth (TSB) - DIFCO 0370-15-5, or the equivalent. PRD1 bacteriophage was received from Dr. Mark Sobsey at the University of North Carolina,

Chapel Hill. The *Salmonella choleraesuis* – ATCC # 19585 was utilized as the host bacterium, with the bacterial growth media being TSB – DIFCO 0370-15-5, or the equivalent.

Media Preparation

Tryptic Soy Agar (Bottom Agar Petri Plates 100 × 15mm) Preparation. The media was rehydrated according to label directions. A magnetic stir bar was placed into the dehydration flask, and the media was brought to a near-boil to dissolve the agar. The flask and contents were then sterilized by autoclaving for 15 minutes after which it was cooled in a water bath to between 45 and 50°C. Plates were poured using approximately 12 to 15 mL per plate. Enough agar was added to cover about two-thirds of the area of the plate. After pouring one plate, the lid was replaced on the dish and gently swirled so that the agar covered the entire bottom of the plate. Plates were allowed to remain motionless until the agar hardened (usually 10 to 15 minutes). Plates were stored at 4°C up to 30 days.

Tryptic Soy Agar Overlay Tubes. The media was rehydrated according to label directions. A magnetic stir bar was placed into the rehydration flask, and the media was brought to a near-boil to dissolve the agar. The media was sterilized for 15 minutes and then pipetted aseptically into 15-mL tubes (3 mL per tube), or pipetted into the tubes (3 mL per tube) which were then capped loosely and sterilized for 15 minutes. The caps were tightened after cooling. Overlay tubes were stored at 4°C for 30 days.

Preparation of High-Titer Bacteriophage

To propagate bacteriophage, a bacterial host slant specific for the bacteriophage of interest was washed with 3 mL of sterile TSB. The total 3 mL was then transferred to a 1-liter flask containing 200 mL of sterile TSB and incubated at 37°C for approximately 3 hours. At this time, the flask was removed from the incubator, 2 mL of bacteriophage stock was added, and then the flask was placed back in the incubator for an additional 24 hours. The bacteriophage/bacteria suspension was poured into four 50-mL centrifuge tubes and centrifuged at 4,000 times gravity for 30 minutes.

During preparation of high-titer bacteriophage stocks, there was a potential for aggregate formation. To reduce aggregates, the bacteriophage stock was filtered through sequentially smaller (0.45-micron and 0.22-micron), low protein-binding filters. To reduce bacteriophage binding to the filter, each filter was pretreated by filtering 10 mL of 0.1% Tween 80 followed by 10 mL of reagent grade non-chlorinated water. Bacteriophage stock preparations were filtered through these pretreated filters with careful attention focused on amount of pressure/vacuum applied to prevent membrane filter failure. Multiple filters were sometimes necessary to filter the entire stock solution. The bacteriophage stock was then titered to determine its concentration and stored at 4°C prior to shipment to the testing facility.

Preparation of Host Culture

The host culture was started the day before the assay was to be performed. Using a sterile swab, a small amount of bacterial host (specific for the bacteriophage of interest) was removed from an agar slant and placed into a sterile tube containing 3 mL of TSB and grown at 37°C for 24 hours. The next day, 1 mL of the overnight culture was pipetted into 50 mL of TSB in a 250-mL

Erlenmeyer flask or the equivalent. The culture was then placed in a 37°C incubator for 4 hours, after which it was removed and placed on ice until used.

Soft Agar Overlay Method for Bacteriophage

Overlay tubes were melted in a boiling water bath or autoclaved for 5 minutes and placed in a 49°C water bath until used. The bottoms of the petri plates were labeled with the identification of each sample to be analyzed. Then 0.1 to 1 mL of the 4-hour host culture (which was in log phase of growth) was pipetted into a pre-warmed overlay tube along with 0.1 to 1 mL of the sample to be analyzed. The tube was mixed by rapidly rolling between the analyst's palms and poured onto a TSB plate. The sample was spread evenly over the surface of the plate by gently and quickly swirling the plate. The plate, which solidified within 30 seconds, was then inverted and incubated at 37°C for approximately 24 ± 2 hours. During the incubation time, the host bacteria formed a confluent lawn over the surface of the petri plate and phage particles that were present in the sample attached to and infiltrated the bacterial host cells. The bacteriophage replicated within the bacterial cells and reached a concentration that lysed (burst) the bacterium. The destruction of the bacterial cells that made up the confluent lawn resulted in clear areas known as plaques. The concentration of bacteriophage originally present in the sample were determined by visually counting the clear areas, which were reported number of plaque forming units per mL (PFU/mL).

Samples were stored at 4°C for no more than 24 hours before the assay. Positive controls were conducted using MS2 bacteriophage stocks of known concentration. Negative controls consisted of phosphate buffered saline (PBS) as the sample media.

Bacteria

Pseudomonas diminuta

Pseudomonas diminuta (ATCC #19146) is a gram-negative small, rod-shaped organism about 0.3 to 0.4 µm by 0.6 to 1.0 µm in size, occurring primarily as single cells. Colonies of *P. diminuta* are yellow-beige, slightly convex, complete and shiny. The initial culture was prepared by reconstituting the culture according to direction provided by the ATCC. Stock cultures were prepared from single cell isolates of *P. diminuta* grown in TSB (Difco Laboratory) at 37 ± 2°C for 24 hours. High-concentration bacterial cultures were prepared by inoculating a single colony of *P. diminuta* into 4 mL of TSB and incubating overnight at 37 ± 2°C for 24 hours. This culture was then used to inoculate a 500-mL volume of sterile TSB that was placed in a shaker incubator at 37 ± 2°C for 24 hours. After this time, the culture was centrifuged for 10 minutes at 3000 × g to pellet the bacterial cells. The supernatant was removed and the bacterial pellet was resuspended in a total of 200 mL of phosphate buffered saline (PBS). This was then divided into four 50-mL aliquots and packaged with blue ice for overnight shipment to the pilot testing facility. The *P. diminuta* cultures were used with 24 hours of receipt at the pilot facility. A shipping control consisting of a sub-sample of the bacterial culture was included with the influent and effluent samples for all phases of challenge testing that employed *P. diminuta*.

All samples of *P. diminuta* were assayed within 24 hours of receipt from the pilot facility. Dilutions of samples were performed using PBS (APHA 1998). Sample analysis was performed using membrane filtration techniques on tryptic soy agar plates (APHA 1998). Solidified agar plates were stored in the refrigerator for a maximum of 2 weeks. Sample plates were incubated at

30 ± 2°C for 48 hours. Colonies on plates showing between 20 and 200 colonies per membrane filter were used to calculate the concentration (CFU/mL) of the original suspensions. Negative control samples consisted of sterile PBS.

Pseudomonas agar was used as an alternative to nutrient agar for assay of *P. diminuta* for samples from the UK pilot testing. The method involved filtering 10 mL of sample through a 0.45-μm membrane filter and then placing the membrane onto *Pseudomonas* agar. The agar was made by adding 48.4 g *Pseudomonas* agar base (Lab MTM; Bury, Lancashire, UK) to 1 L of deionized water followed by the addition of 10 mL of glycerol. The agar was sterilized by autoclaving at 121°C for 15 minutes. The agar was allowed to cool to 47°C before the addition of two vials of 108 C.F.C. supplement (Lab MTM; Bury, Lancashire, UK).

Escherichia coli

E. coli (ATCC #15597) is a gram-negative, non-spore forming rod-shaped bacteria that produce gas upon lactose fermentation in prescribed culture media within 48 hours. Colonies of *E. coli* appear small, round and possess a characteristic metallic green sheen when grown on m-Endo agar. The initial culture was prepared by reconstituting the culture according to direction provided by the ATCC. Stock cultures were prepared from single-cell isolates of *E. coli* grown in TSB (Difco Laboratory) at 37 ± 2°C for 24 hours. High-concentration bacterial cultures were prepared by inoculating a single colony of *E. coli* into 4 mL of TSB and incubating overnight at 37 ± 2°C for 24 hours. This culture was then used to inoculate a 500-mL volume of sterile TSB that was placed in a shaker incubator at 37 ± 2°C for 24 hours. After this time, the culture was centrifuged for 10 minutes at 3000 × g to pellet the bacterial cells. The supernatant was removed and the bacterial pellet was resuspended in a total of 200 mL of PBS. This was then divided into four 50-mL aliquots and packaged with blue ice for overnight shipment to the pilot testing facility. The *E.coli* cultures were used with 24 hours of receipt at the pilot facility. A shipping control consisting of a sub-sample of the bacterial culture was included with the influent and effluent samples for all microbial challenge testing.

All samples of *E. coli* were assayed within 24 hours of receipt. Dilutions of samples were performed using PBS. *E. coli* bacteria were analyzed by membrane filtration using m-Endo agar plates according to manufacturer's specifications. Solidified agar plates were wrapped in tin foil and stored in the refrigerator for a maximum of two weeks. After samples had been passed through membrane filters, each filter was placed on an m-Endo media plate and incubated, inverted, for 24 ± 2 hours at 37 ± 2°C. After incubation, colonies were identified as *E. coli* by their characteristic metallic green sheen. A negative control sample was prepared by filtering sterile PBS through a sterile membrane, placing it on m-Endo and incubating as described above.

Klebsiella terrigena

Klebsiella terrigena is a gram-negative, non-spore forming rod-shaped bacteria that produce gas upon lactose fermentation in prescribed culture media within 48 hours. The initial culture was prepared by reconstituting the culture according to direction provided by the ATCC. Stock cultures were prepared from single-cell isolates of *K. terrigena* grown in TSB (Difco Laboratory) at 37 ± 2°C for 24 hours. High-concentration bacterial cultures were prepared by inoculating a single colony of *K. terrigena* into 4 mL of TSB and incubating overnight at 37 ± 2°C for 24 hours. This culture was then used to inoculate a 500-mL volume of sterile TSB that was placed

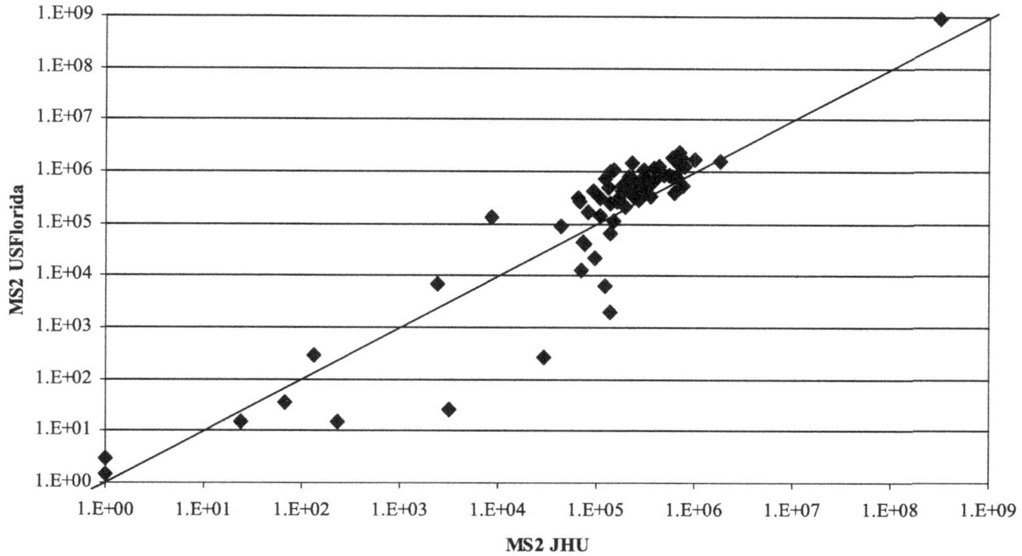

Figure 2.15 Inter-laboratory comparison of MS2 bacteriophage enumeration

in a shaker incubator at 37 ± 2°C for 24 hours. After that time, the culture was centrifuged for 10 minutes at 3000 × g to pellet the bacterial cells. The supernatant was removed and the bacterial pellet was resuspended in a total of 200 mL of PBS. This was then divided into four 50-mL aliquots and packaged with blue ice for overnight shipment to the pilot testing facility. The *K. terrigena* cultures were used with 24 hours of receipt at the pilot facility. A shipping control consisting of a sub-sample of the bacterial culture was included with the influent and effluent samples for all microbial challenge testing that used *K. terrigena.*

All samples of *K. terrigena* were assayed within 24 hours of receipt. Dilutions of samples were performed using PBS. *K. terrigena* bacteria were analyzed by membrane filtration using m-Endo agar plates according to manufacturer's specifications. Solidified agar plates were stored wrapped in tin foil in the refrigerator for a maximum of two weeks. After samples had been passed through 0.45-μm membrane filters, each filter was placed on an m-Endo media plate and incubated, inverted, for 24 ± 2 hours at 37 ± 2°C. After incubation, colonies were verified as *K. terrigena* by their characteristic metallic green sheen. A negative control sample was prepared by filtering sterile PBS through a sterile membrane, placing it on m-Endo and incubating as described above.

Quality Assurance/Quality Control

An initial step in the microbial removal experiments was to verify quality control in the MS2 bacteriophage assay between the two US analytical laboratories (JHU and USF). Split samples of MS2 bacteriophage were assayed at JHU and USF. Figure 2.15 shows the variation around a line of perfect agreement. In general, the MS2 bacteriophage data were consistent between the two laboratories. Variation was normally within one log of the line. No experimental bias was observed since there was equal scatter on either side of the agreement line.

Animal Viruses

Cell Culture Protocols

Growth Media Preparation. The following materials were required: 1X Minimal Essential Medium (MEM) with phenol red, fetal calf serum, newborn calf serum, sodium bicarbonate, HEPES buffer, non-essential amino acids, L-glutamine, gentamicin and kanamycin, autoclave bin, 10-ml pipettes, two graduated cylinders (50 or 100 ml), water bath set at 37°C, water bath set at 56°C, and hypochlorite solution.

The media requirements were estimated for a week's passages and production of all cell lines. Allowance was made for 200 to 300 mL of extra media. Three 150 cm^2 flasks and a minimum of 1 roller bottle (RB) were generally made for cell lines which had the highest demand in production for hold-backs and passage. Two 150-cm^2 flasks were prepared for other cell lines.

Fetal calf serum, newborn calf serum, L-glutamine, gentamicin/kanamycin, and non-essential amino acids were removed from the freezer, and thawed in the 37°C water bath. HEPES buffer, sodium bicarbonate, and 1X MEM were kept in the refrigerator at 4°C. 1X MEM was labeled with the date and the type and concentration of calf serum which was added. Fetal calf serum was used for passages. Newborn calf serum was used for production. Most cell lines required 10% serum per volume for growth; e.g., 10 mL of serum per 100 mL of MEM. FRhK-4 cells required 15% serum per volume for growth; e.g., 15 mL of serum per 100 mL of MEM. When the fetal calf and newborn calf sera were completely thawed, they were heat-inactivated for 30 minutes at 56°C. Certain proteins, inhibitory to cell proliferation, were denatured by this heating process.

Initially, the hood was turned on and disinfected with 70% ethanol. The tops of all bottles of solutions were flamed and placed in the hood. Ten-ml pipettes were used to add to all bottles of 1X MEM:

- 100 mL of 1X MEM
- 1mL of 100X L-glutamine
- 1 mL of 100X non-essential amino acids
- 1 mL of 100X gentamicin/kanamycin
- 1 mL of 100X HEPES buffer
- 1.5 mL of 100X sodium bicarbonate

Serum was then added according to the bottle labels using graduated cylinders. Separate graduated cylinders were used for fetal and newborn calf serum. When finished, the media was a clear, reddish-orange color due to the phenol red pH indicator.

Cell Passage. The following materials were required: trypsin-EDTA, PBS, growth media previously prepared with fetal calf serum, graduated cylinder (50 or 100 mL), 10 mL pipettes, flasks and RBs, sterile discard beaker, 50-mL centrifuge tubes, table-top centrifuge (non-refrigerated), balance tubes, Bunsen burner and striker, water bath set at 37°C, hypochlorite solution and autoclave bin.

Trypsin-EDTA was thawed, and the media and PBS were warmed in a 37°C water bath. The sterile hood was turned on at least 15 minutes before it was used. The bench was disinfected twice with 70% ethanol. The tops of all bottles, flasks and centrifuge tubes were flamed before being placed into the hood. A 1:10 dilution was made of 10X EDTA-trypsin in PBS, 10 mL:90 mL. Media was dispensed into flasks and roller bottles for passage using the graduated cylinder:

- 25-cm^2 flask: 10 mL
- 75-cm^2 flask: 25 mL
- 150-cm^2 flask: 50 mL
- 900-cm^2 RB: 150 mL

25 to 50 mL of media was kept in the bottom of one bottle for resuspending the cell pellets. Two mL of media was added to each centrifuge tube to keep the trypsin from digesting cell membranes. Cells were removed from the walk-in incubator and examined microscopically to verify their health. Hold-back flasks of cells were kept for 3 consecutive weeks. The cell passage from the previous week was used for the current week's passage. The oldest cells were put in the autoclave bin to be autoclaved and discarded. The caps of the flasks/RBs for the current week's passage were flamed and placed in the hood. Old media was poured into the discard beaker. The flask was rinsed with PBS to remove any residual media that could interfere with the trypsinization of the cell layer:

- 25 cm^2 flask: 5 mL
- 75 cm^2 flask: 7 mL
- 150 cm^2 flask: 10 mL
- 175 cm^2 flask: 12 mL
- 900 cm^2 RB: 20 mL

The PBS was decanted into the waste beaker. The trypsin/PBS solution was added to the flask or RB using the same amount as in the rinsing step. The flask was placed in the walk-in incubator at 37°C. The amount of time needed to completely trypsinize the cell layer varied with each cell line. The flasks were checked every 2 minutes. Trypsinization was complete when the cell layer appeared hazy and as if it had been scratched. Agitation was used as necessary to loosen the cells from the flask. The flasks were reflamed and the trypsinized cells were aseptically poured into the centrifuge tubes prepared earlier. The cells were centrifuged for 8 to 10 minutes at 2,000 rpm. The centrifuge tubes were flamed and the supernatant was decanted into the waste beaker under the hood. The pellet was resuspended in the appropriate volume of media using a 10-mL pipette. The cells were split in the proper ratio into the prepared flasks or RB, the caps were tightened and the container was gently swirled to disperse cells evenly. The container was labeled with cell line, passage level and date, and placed in the walk-in incubator. Information was recorded in the cell passage log book.

Cell Production. The following materials were needed: hemocytometer and cover slip, hand-held counter, Wheaton siphon pump, stir plate and sterile stir bar, sterile tubing and holder, sterile beaker or flask, growth media previously prepared with calf serum, dilution tubes, 1-mL and 10-mL pipettes, hypochlorite solution, 50-mL graduated cylinder, trypsinized cells, sixty 15-mm sterile tissue culture dishes, and flasks or RBs.

The cells for production were trypsinized the same way as for passage and were usually prepared at the same time as the passage. To make calculating cell numbers easier, the pellet was usually resuspended in 10 mL of media. A 1:10 dilution of the resuspended cells (cell stock) was made by adding 0.1 mL of cell suspension to 0.9 mL of medium and mixing well. A second or third serial dilution was sometimes necessary depending on how thick the cells grew; each cell line grows differently. The number of dilutions also depended on the size of the container in which the cells were grown. The hemocytometer and cover-slip were stored in alcohol, then wiped dry. A small

Table 2.12
Total media volume required for cell production

Container	Media volume
96-well plate	0.5 mL
24-well plate	1 to 2 mL
6-well plate	2 to 3 mL
Sixty 15-mm dishes	5 mL
25 cm^2 flask	10 mL
75 cm^2 flask	25 mL
150 cm^2 flask	50 mL
900 cm^2 roller bottle	150 mL

amount of the final dilution was loaded into both chambers of the hemocytometer and cells were counted as soon as possible. Cells were not allowed to stand for more than 30 minutes to avoid the formation of clumps of cells that make counting difficult and can depress cell counts. The cell numbers were determined by counting the cells in the four corner blocks of each chamber (each corner block was divided into 16 squares). In total, 8 blocks were counted and the total number of cells was divided by 8. The volume of each 16-square block was 1×10^{-4} mL. The number was multiplied by 10,000 to calculate cells per mL at a dilution; this number was then multiplied by the dilution factor (ratio) for the cells per mL of stock cells. To calculate the number of cells harvested from each culture vessel, the number of cells per mL was multiplied by the volume used to resuspend the pellet. Once the number of cells per flask or RB was determined, the number of flasks or RBs needed to complete the production could be determined as shown in Table 2.12.

Duplicate dilutions and counts were made and counts were repeated unless within 15%. Stainless metal trays were cleaned with 70% ethanol and labeled with cell line, date and name or initials of the person who ordered the cells. Dishes on trays while checking for breaks or cracks. The stir plate was placed in the hood. Warmed media bottles were flamed and placed in the hood. Media was poured into the beaker or flask, allowing 50 to 75 mL of media extra for production. The stir bar was added. The appropriate number of cells were added to the media, ensuring that the mixture was stirred well throughout dispensing to avoid the cells settling and dispensing unevenly. The mixture was stirred for approximately 5 minutes before dispensing. The Wheaton pump was set up under the hood. Sterile tubing was run from the dispensing beaker through the pump, with both ends of the tubing remaining sterile. The nozzle end of the tubing was clamped in the holder and the other end was carefully lowered into the media and cell solution until it touched the bottom of the beaker. The pump was calibrated using the graduated cylinder, and the cells were dispensed. Cells in dishes were placed in the CO_2 incubator with 0.5% CO_2 at 37°C. Cells in flasks or RBs were placed in the incubator.

Viral Cell Assays as Represented by Poliovirus Assay

Poliovirus type 1, strain LSc, were propagated in BGMK (African green-monkey kidney derived) cells and assayed via the plaque technique in confluent layers of BGMK cells. Between 10^7 and 10^8 plaque forming units per mL of viral stocks were generated.

Poliovirus Propagation. BGMK cells were prepared in RBs or tissue culture flasks (T-75s or T-150s). Each T-150 contained approximately 3×10^7 cells, each T-75 contained approximately 1.5×10^7 cells and each RB contained approximately 2×10^8 cells. When cells were about 3 to 4 days old, or when there was >90% confluency, the cells were ready for use. Inoculating cells just before 100% confluency was the preferred practice.

Using sterile technique, growth media was poured off from the RB, and rinsed with maintenance media (see protocol below). The rinse was decanted. The seed stock of poliovirus was inoculated onto newly confluent BGMK cell layers at a multiplicity of infection of approximately 0.1 pfu/cell. An inoculum volume of approximately 1 to 2 mL was used for each T-150 or approximately 5 to 10 mL for each RB (of virus in PBS or serum-free maintenance media). The virus was allowed to adsorb to cells by incubating RBs at 37°C for 30 to 60 minutes. 70 mL of serum-free maintenance media was added to each RB (30 to 50 mL per T-150, or 20 mL per T-75). Total volume in each RB was approximately 80 mL at this point. Infected cells were incubated for 3 to 5 days at 37°C or until the observed cytopathic effect on the monolayer was >95%.

Harvest of poliovirus from infected cells was begun by freezing and thawing three times. The freezing was performed at -20°C and the thawing was performed in the incubator at 37°C. For the freezing steps, the bottles or flasks were allowed to freeze solidly (approximately 1 to 2 hours minimum). The cell lysate was homogenized with one-half to an equal volume of freon (1,1,2-trichloro-1,2,2-trifluoroethane) or chloroform at high speed for 2 minutes with an Omni Mixer (Omni International, Waterbury, CT). The homogenate was centrifuged at 4,000 to 5,000 × g for 15 minutes. The supernatant was recovered as a virus stock, aliquotted and frozen at -80°C.

Poliovirus and Other Virus Monodispersion. Virus stocks were monodispersed by passage through 0.2- and 0.08-µm Nucleopore membrane filters that had been pre-treated with 0.01% Tween-80 to reduce viral adsorption to the filters. Monodispersed viral stocks were ready for seeding experiments.

Poliovirus Assay. Spent media was aspirated into a filter flask. 100 to 200 µL of sample was inoculated onto BGMK monolayers in 60-mm dishes and incubated at 37°C in a CO_2 atmosphere incubator. Dishes were incubated for 60 minutes with gentle rocking every 20 minutes to prevent the monolayer from drying out. Negative controls were always included.

Infected monolayers were overlayed with a 1% agarose overlay medium (see protocol for agarose overlay) and cultures were incubated at 37°C for 3 to 5 days. Cultures were checked daily for fungal contamination. Plaques were counted and recorded as plaque forming units per mL.

Agarose Overlay Materials. The appropriate volume of 1% agarose (electrophoresis grade) was prepared by adding 1 g of agarose to 100 mL of distilled, deionized water in a 500-mL bottle. The bottle was autoclaved and placed in a 45 to 50°C water bath immediately. The agarose was placed into the 45°C water bath 1 to 2 hours before the assay.

100-mL portions of 2X MEM were aliquotted into 125-mL bottles and placed in a 37°C water bath 1 to 2 hours before the assay. Generally 100 mL of 2X MEM plus 100 mL of 1% agarose plus 1 tube of supplements (see below) was enough for approximately 20 dishes or 1 large aluminum tray of dishes. Frozen media supplements were removed from the -20°C freezer and placed in a 37°C water bath to thaw 1 to 2 hours before the assay. Each tube contained 18.5 mL. When the dishes of BGMK cells were ready to receive overlay, one 100-mL portion of 2X MEM and one 18.5-mL portion of supplement were removed from the water bath and the lids were flamed. The thawed supplements were vortexed well. One tube (18.5 mL) of supplements was added to 100 mL of 2X MEM.

Table 2.13
Media supplements for agarose overlays and maintenance (for 1L bottle)

Reagent	1X volume (mL)	50X volume (mL)
7.5% $NaHCO_3$	1.5	75
HEPES buffer	1.0	50
L-glutamine	1.0	50
Nonessential amino acids	1.0	50
100X gentamicin/kanamycin	1.0	50
Nystatin	1.0	50
4M $MgCl_2$	0.75	37.5
Fetal calf serum	2.0	100
Total	**9.25**	**462.5**

The agarose was removed from the water bath and the lid was flamed. The 118.5-mL volume of 2X MEM and supplements was added to the agarose and swirled gently. The agarose mixture was then added to the dishes using a 25-mL pipette. Approximately 7 mL was pipetted to the side of each dish, ensuring that the agarose material was not so hot that it could damage cells. The agarose overlay mixture was allowed approximately 5 to 10 minutes to solidify before replacing the dishes in the incubator.

Reagents. Media supplements for agarose overlays and maintenance were prepared according to Table 2.13.

Frozen reagents were placed in a 37°C water bath to thaw. When completely thawed, the reagents were removed from the water bath and lids were flamed. Volumes of reagents were added to a 1-L bottle as described in Table 2.13. The bottle was mixed well by inverting the bottle several times with foaming minimized. 462.5 mL was divided into 18.5-mL portions in 50-mL centrifuge tubes, for approximately 25 aliquots. The centrifuge tubers were labeled and stored at -20°C. One tube was used per 100 mL of agarose, and 100 mL of agarose per 100 mL of 2X MEM. 100 mL of water was added to 81.3 g $MgCl_2$-$6H_2O$ in a large beaker, with the water added incrementally. The mixture was autoclaved and stored at 4°C.

Maintenance Media. 1X MEM was used for maintenance media. One centrifuge tube of thawed supplements was added to 200 mL of 1X MEM. Maintenance media was used for propagating virus. After the cells had been inoculated and allowed to infect for 1 hour, the appropriate volume of maintenance media was added to the RB or flask. The maintenance media was warmed to 37°C before inoculating.

Poliovirus maintenance media was comprised of:

- 100 mL of 1X MEM
- 1 mL of HEPES buffer
- 1.5 mL of 7.5% $NaHCO_3$
- 1 mL of 100X gentamicin/kanamycin
- 0.7 mM of $MgCl_2$ (added only for virus propagation).

CHAPTER 3
BENCH-SCALE PROTOCOL FOR CHARACTERIZATION OF MEMBRANES USING MICROBIAL PROFILING

INTRODUCTION

Microbial removal is usually evaluated through pilot testing. However, rigorous microbial challenge studies at pilot scale are often prohibitively costly. Since pilot studies are typically conducted at water facilities, opportunities for microbial challenge studies may be limited due to the potential hazard of working with microorganisms in proximity to drinking water supplies. Further, membranes are most vulnerable to microbial passage when they are first put into service. There are numerous complexities of sampling for microbial agents immediately after pilot plant start-up (such as collecting a sample rapidly and aseptically). The inability to sample a clean membrane as soon as it is placed on line may provide inconsistent results due to deposition layer accumulation or pore constriction from adsorption of natural organic matter onto the membrane.

This chapter provides a protocol to evaluate microbial removal by membranes at bench scale. Materials for most low-pressure membrane filtration modules used in drinking water treatment have a hollow fiber geometry. Nonetheless, this protocol can be employed for other geometries (i.e., tubular) with little modification. The intent of the protocol is to provide a widely accepted, standardized methodology with which to characterize membranes from a microbial perspective. The absence of natural water constituents in the feedwater will allow accurate assessment of the capability of membrane materials to remove microorganisms.

The protocol is designed for scientific rigor, but also for ease of implementation by a qualified laboratory, denoted as the "Laboratory Testing Organization" throughout this chapter. The data generated from execution of the protocol are intended to provide utilities, engineers and regulators with the necessary information to make informed decisions about current and future membrane products that are being (or will be) employed for treatment of water supplies. This protocol may also be used by the manufacturer to determine the membrane lot acceptance of a membrane filtration media when challenge testing is needed to demonstrate microbial removal for lot acceptance.

It is important to note that this chapter is written as a stand-alone document with the intent of being able to use it for microbial characterization of membranes without reading the report in its entirety. It should also be noted that this protocol may be used for verification under the Environmental Testing Verification (ETV) Program under the National Sanitation Foundation (NSF) and the USEPA. If seeking verification under this program, modifications to the protocol shall be reviewed and approved by the NSF and EPA under the ETV Program prior to implementation.

GENERAL APPROACH

The general approach to the membrane bench testing is to conduct microbial challenge studies under conditions in which the microorganisms are most likely to penetrate the membrane using a standardized Low-Pressure Membrane Testing Unit. The primary operational variable shall be transmembrane pressure, which shall be applied under direct flow conditions. Feedwater containing the selected microorganisms shall be applied continuously to the membrane for the duration of the challenge. No backwashing or recirculation shall be employed during the experimental run. The challenge studies shall be executed through a series of tasks as noted below and discussed in more detail after a discussion of the membrane testing unit experimental set-up.

- Task A: Establish Filtrate Flux
- Task B: Perform Membrane Cleaning and Backwashing
- Task C: Perform Membrane Integrity Testing
- Task D: Conduct Microbial Removal Experiments
- Task E: Execute Data Handling Protocol

MEMBRANE FILTRATION UNIT EXPERIMENTAL SET-UP

Low-Pressure Membrane Testing Unit

The Low-Pressure Membrane Testing Unit in this document refers to the membrane module, associated tubing and connections, pressure gauges, tanks and pumps (or nitrogen tanks). The unit can be set up in three different ways, depending on the type of flow configuration necessary for the particular membrane module to be tested. These three different experimental set-ups for the Low-Pressure Membrane Testing Unit are illustrated in Figures 3.1 through 3.3. The three set-ups accommodate two pressure-driven flow configurations (inside out and outside in) as well as one outside-in, vacuum driven configuration. The same Low-Pressure Membrane Testing Unit can be employed for each flow configuration with minor tubing changes.

The Low-Pressure Membrane Testing Unit is comprised of five separate tanks: one tank provides a microbial feed for the challenge studies; three other tanks (chlorine or other biocidal agent, sodium thiosulfate and phosphate buffer) are employed for backwashing and/or cleaning the membrane module between experimental runs. A fifth tank is used to collect waste. Pressurized nitrogen gas is employed to provide system pressure, since mechanical pumping can cause perturbations in pressure application. A vacuum system is employed for submersible vacuum driven applications. Details of materials used for construction of the Low-Pressure Membrane Testing Unit are presented in Appendix A.

Bench-Scale Modules Used in Challenge Testing

Membranes exhibit product variability, and the degree of variability will depend on a number of factors specific to the product and the manufacturing process. Furthermore, product variability is manifested in different aspects of product performance and characteristics. Membranes may exhibit variability with respect to removal efficiency, pore size distribution, bubble point, productivity, and membrane area among other characteristics. All of these aspects of product variability can be important considerations for a specific application; however, variability in removal efficiency is of primary concern in the context of challenge testing.

The membrane material used by the manufacturer to fabricate the bench modules will be obtained from full-scale modules or membrane materials from their production line. The membrane modules used for evaluation in the Low-Pressure Membrane Testing Unit shall be based on the statistical distribution of nondestructive performance test results, as described below, or on an alternative approach provided by the manufacturer and Laboratory Testing Organization. (If seeking verification under the ETV Program and an alternative approach is desired, the alternative

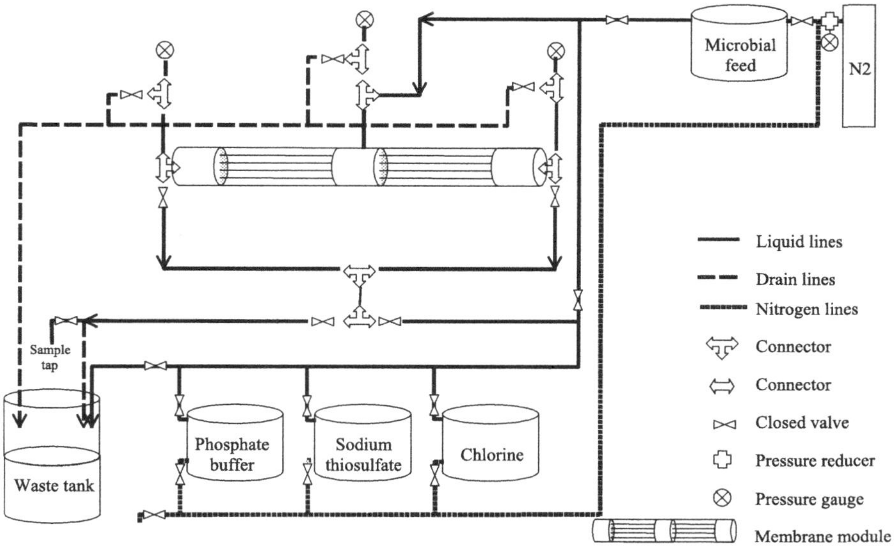

Figure 3.1 Schematic of bench-scale low pressure membrane testing unit (outside/in flow configuration, pressure-driven)

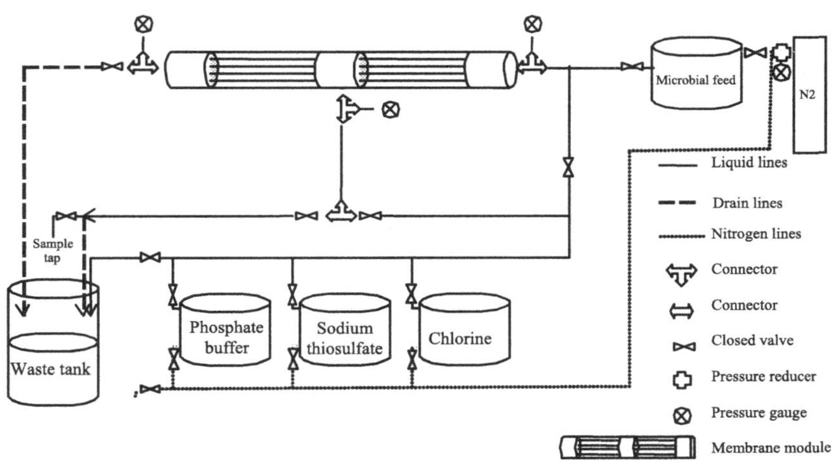

Figure 3.2 Schematic of bench-scale low pressure membrane testing unit (inside/out flow configuration, pressure driven)

approach shall be reviewed and approved by NSF and the EPA under the ETV Program prior to implementation.) Alternative approaches shall take into account product variability in terms of efficacy of microbial removal and may depend upon the specific characteristics of the membrane.

Non-Destructive Membrane Test and Quality Control Release Values

A nondestructive performance test is a physical test that characterizes a fundamental property of the membrane that can be correlated to some aspect of process performance, and which

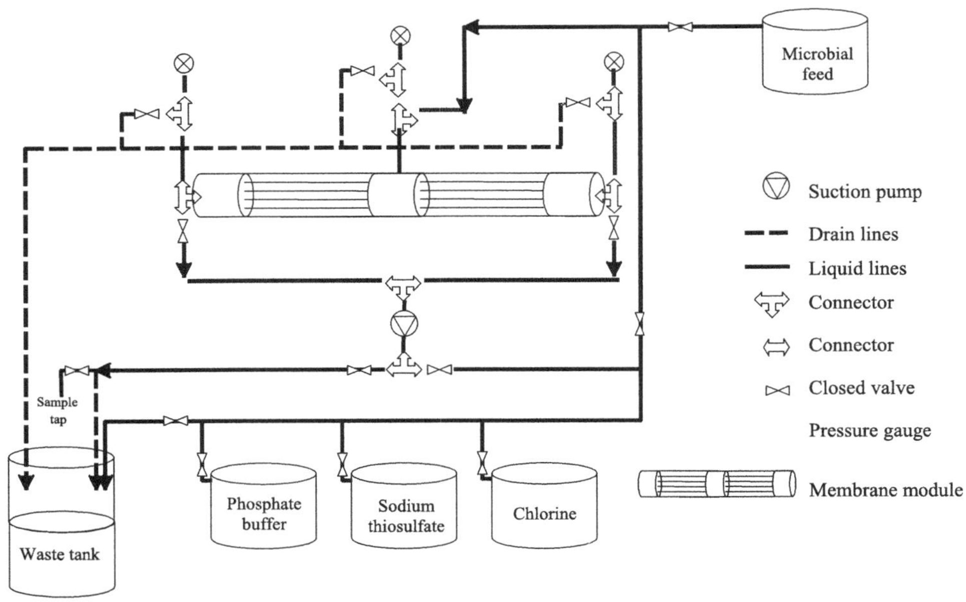

Figure 3.3 Schematic of bench-scale low pressure membrane testing unit (outside/in flow configuration, vacuum driven)

does not alter or damage the membrane. In the context of challenge testing, the nondestructive performance test must be correlated to the removal efficiency of the target organism. An example of such a test is the bubble point test, the results of which can be directly related to the size of the largest pore in a membrane. Manufacturers often use such nondestructive testing as a means of quality control and assurance, and in many cases such a test is applied to every production module. The results of such extensive testing can provide a good estimate of product variability as it relates to removal efficiency.

The nondestructive performance test is used to assure the removal efficiency of production modules in the following manner. The microbial challenge test demonstrates the removal efficiency of the specific module(s) evaluated, and these modules are characterized through application of the nondestructive performance test that is used as part of the manufacturer's routine quality control and assurance program. The results of the nondestructive performance test applied to the module(s) evaluated during the challenge test establish a control limit for the nondestructive performance test.

The nondestructive performance serves as the basis for confirming the performance of membrane modules that are not directly challenge tested with microorganisms. Manufacturers that have historically performed nondestructive testing for the purpose of product quality control and assurance can use this information to characterize the variability of a product line. Additionally, a manufacturer may have established a "quality control release value" for the nondestructive performance test that provides a minimum cut-off for an acceptable product. Based on these considerations, selection of membrane materials used in bench-scale modules for microbial challenge testing should be based on nondestructive performance test results.

Since the microbial challenge test is used to establish the control limit for the nondestructive performance test that production modules must meet, it would be prudent to test modules with nondestructive performance test results that are close to the quality control release value. The rationale behind testing a worst-case module is that it allows for fewer modules to be tested while still providing a means of verifying the removal efficiency of production modules through application of the nondestructive performance test. This approach for module selection may be especially useful when complete removal of the challenge organism is anticipated for all modules of the product line.

It should be noted that many nondestructive performance tests that are suitable for evaluating the ability of a membrane to remove large organisms, such as *Cryptosporidium,* will not apply to very small ones, such as viruses. As an example, the bubble point test cannot typically be applied at a pressure high enough to achieve a resolution on the order of the size of a virus. In such cases, other manufacturing quality control procedures would be necessary to assure virus removal capabilities of production modules. These may include internal quality control testing of the membrane media or testing for acceptance of membrane modules. For more detailed information regarding non-destructive membrane tests and quality control release values, USEPA *Membrane Filtration Guidance Manual* (USEPA, 2003b) should be consulted.

Bench-Scale Membrane Module Requirements

Manufacturers shall construct modules with materials as close as possible to their quality control release. Appropriate information on the nondestructive performance tests for these modules or other quality control information shall be provided. A minimum of five bench modules to be employed for the each bench-scale microbial characterization shall be provided by the manufacturer. Each module shall have an effective membrane area of approximately 0.1 m^2. The modules shall have an effective length similar to those employed at full scale in order to have similar pressure drops along the membrane. It is recognized that the length of the bench-scale module may be slighter shorter than that of a full-scale module if the materials to fabricate it are taken directly from a full-scale module. Further, it is recognized that the geometry of some tubular and spiral wound membranes may preclude the use of full-scale length modules or elements. In these cases, the length necessary to provide 0.1 m^2 modules should be employed.

Conditioning of Membrane

Before conducting testing, the membranes shall be fully wetted according the manufacturer's specification. After wetting, each module shall be conditioned at a typical filtrate flux (as specified by the manufacturer) for a minimum of 4 continuous hours before any testing begins. A 0.1 mM phosphate buffer solution (pH 7.0) shall be employed as the feedwater. Specific flux shall be monitored once per hour and recorded.

SEQUENCE OF EVENTS FOR MODULE TESTING

Table 3.1 presents the general sequence of events for module testing. These events are described in more detail in each of the tasks.

Table 3.1
Sequence of events for low-pressure membrane module testing

Event	Comments
Conditioning period for module	Run module for 4 hours
Perform first membrane integrity test	Described in Task C
Set filtrate flux; determine specific flux	Conduct before and after five HRTs; chemically clean only if necessary; described in Task A
Perform microbial challenge test on module	Described in Task D
Determine specific flux	Described in Task A
Perform second membrane integrity test	Described in Task C
Repeat same sequence with other modules	

TASK A: ESTABLISH FILTRATE FLUX

Introduction

Bench-scale membrane operation in terms of filtrate flux will be established in this task. This task shall be conducted for each of the five membrane modules being tested. This task shall also be repeated after each chemical cleaning (see Task B), if chemical cleaning is necessary. Each repetition of this task involves filtration of 0.1 mM phosphate buffer solution for five HRTs of the low-pressure membrane test unit's module and tubing at a filtrate flux under which microorganisms would be most likely to penetrate the membrane.

Experimental Objectives

The objectives of this task are to document the operational conditions under which each of the five membrane modules will be evaluated for microbial removal and then to verify those operational conditions before and after testing of each membrane module. While five modules are the minimum to be evaluated, the testing of more modules is encouraged.

Work Plan

Specification of Filtrate Flux

For this task, the Laboratory Testing Organization shall specify the filtrate flux to be employed during the microbial challenge studies. The microbial challenge experiments shall be conducted at operating conditions under which the microorganisms would be most likely to penetrate the membrane. These conditions shall include the highest operational flux specified by the manufacturer for their membrane using a 0.1 mM phosphate buffer solution and direct flow hydraulic conditions. Note that the pressures to obtain the highest operational flux using a phosphate buffer feedwater may not be as high as those observed under field applications. This is because materials in natural water foul the membrane over time, and thus, greater pressures are

Figure 3.4 Low pressure membrane testing unit

required to maintain a constant flux. However, under these conditions, microbial removals are less conservative than those under the clean water conditions specified here.

It should also be noted that some hollow fiber, and most spiral wound and tubular membranes often operate in the field under crossflow conditions. Nonetheless, direct flow still represents a worse-case scenario since membrane surface concentration polarization effects do not play a role under the conditions in which these experiments are to be conducted. Regardless of membrane geometry, the Laboratory Testing Organization shall clearly describe how operational filtrate flux was selected.

It is anticipated that the filtrate flux will be constant over the time of the experiment, since it is short in duration (see Figure 3.4). However, if greater than 10 percent specific flux decline of the membrane occurs after filtering 0.1 mM phosphate buffer for five HRTs of the low-pressure membrane test unit's module and tubing, chemical cleaning shall be performed according to manufacturer specifications. Adjustments to the operational strategy shall be made (such as a decrease in filtrate flux) as necessary. Decisions on adjustments of filtrate flux shall be made by the Laboratory Testing Organization in consultation with the manufacturer.

Microbial challenge studies at additional operational flux conditions are at the discretion of the manufacturer and the designated Laboratory Testing Organization. However, testing of alternate additional operational conditions shall be performed only in addition to the initial flux condition specified in the Work Plan.

Determination of Specific Flux

On each new module, and before and after each microbial challenge experiment, the hydraulic performance of the membrane module shall be evaluated by determining its specific flux. The required parameters to calculate the specific flux include:

- filtrate flow rate,
- effective membrane area, and
- transmembrane pressure.

To evaluate the filtrate flow rate of the membrane, a volume of filtrate is collected over a period of one minute.

The effective membrane surface area is determined as:

$$A = \pi \times (OD) \times (l) \times (n) \tag{3.1}$$

where: A = effective membrane surface area in m^2 or ft^2,
l = the length of the module in m^2 or ft^2,
OD = outside diameter (OD) of the fibers (for an outside/in flow configuration) in m^2 or ft^2
n = number of fibers.

For an inside/out flow configuration the effective area becomes:

$$A = \pi \times (ID) \times (l) \times (n) \tag{3.2}$$

where: A = effective membrane surface area in m^2 or ft^2,
l = the length of the module in m^2 or ft^2,
ID = inside diameter (ID) of the fibers in m^2 or ft^2, and
n = number of fibers.

The filtrate flux is determined empirically as:

$$J = \frac{Q_p}{A} \tag{3.3}$$

where: J = filtrate flux in $L/hr/m^2$ or $gal/day/ft^2$ (gfd),
Q_p = filtrate flow rate in L/hr or gal/day, and
A = effective membrane surface area in m^2 or ft^2.

In a direct filtration mode, the transmembrane pressure is calculated according to:

$$P_{tm} = P_i - P_p \tag{3.4}$$

where: P_{tm} = transmembrane pressure in bar or psi,
P_i = pressure at the inlet of the module in bar or psi, and
P_p = filtrate pressure in bar or psi.

The water volume transfer through the membrane per unit of membrane area and driving force is the specific flux (J_s) as described by:

$$J_s = \frac{J}{P_{tm}} \tag{3.5}$$

where: J_s = specific flux (permeability) in L/hr/m^2/bar or gfd/psi.

J = filtrate flux in L/hr/m^2 or gfd at 20°C,

P_{tm} = transmembrane pressure in bar or psi.

The filtrate flux is normalized by dividing by the transmembrane pressure or net driving force to obtain the specific flux, which is a useful measure by which different membrane operating conditions can be compared to each other.

Temperature corrections to 20°C for filtrate flux shall be made to correct for the variation of water viscosity with temperature. A specific, empirically derived equation developed by the membrane manufacturer may be used to provide temperature corrections. Alternatively, the following equation by Streeter and Wiley (1985) may be employed:

$$J = \frac{Q_p \times e^{-0.239 \times (T - 20)}}{A} \tag{3.6}$$

where: J = instantaneous flux in L/hr/m^2 or gfd,

Q_p = filtrate flow rate in L/hr or gal/day,

T = temperature in °C, and

A = effective membrane surface area in m^2 or ft^2.

Evaluation Criteria and Minimum Reporting Requirements

- Bar graph of specific flux normalized to 20°C before and after preconditioning for five HRTs and before and after challenge testing of each module.

TASK B: ASSESS CLEANING EFFICIENCY

Introduction

Although not anticipated, chemical cleaning of the membrane may be necessary. As such, this task describes the general procedures for conducting chemical cleaning.

Work Plan

Cleaning chemicals and cleaning routines shall be based on the recommendations of the manufacturer. The manufacturer and the Laboratory Testing Organization shall specify in detail the procedure(s) for chemical cleaning of the membranes. At a minimum, the following shall be specified:

- cleaning chemicals and concentrations,
- hydraulic conditions of cleaning (flow, transmembrane pressure),
- duration of each cleaning step, and
- initial and final temperatures of chemical cleaning solution.

Each system shall be chemically cleaned using the recommended cleaning solutions and procedures specified by the Manufacturer. After each chemical cleaning of the membranes, the system shall be restarted and the initial conditions of specific flux recovery and microbial rejection capabilities shall be tested.

Evaluation Criteria and Minimum Reporting Requirements

At the conclusion of each chemical cleaning, the initial condition of transmembrane pressure, flow and temperature shall be recorded and the specific flux calculated. The efficacy of chemical cleaning shall be evaluated by the recovery of specific flux.

$$\% \text{ recovery of specific flux} = 100*(Js_{ac} / Js_i) \qquad (3.7)$$

where: Js_{ac} = specific flux (L/(hr/m^2/bar or gfd/psi,) after chemical cleaning at end of run
Js_i = initial specific flux (L/(hr/m^2/bar or gfd/psi) at beginning of filtration run

- Table of percent specific flux recoveries before and after each chemical cleaning.

TASK C: PERFORM MEMBRANE INTEGRITY TESTING

Introduction

Monitoring of membrane integrity is necessary to ensure that an adequate barrier is continuously being provided by the membrane material during the challenge testing. Only direct membrane integrity monitoring shall be employed in the bench-scale testing. Examples of direct monitoring methods include, but are not limited to:

- air pressure decay testing,
- diffusive air flow testing,
- water displacement test
- bubble test,
- sonic wave sensing.

A brief overview of these direct monitoring methods is provided below.

Air Pressure Decay Test (PDT)

In this test, the membrane module is pressurized to approximately 1 bar (15 psi) from the feed side. Minimal loss of the held pressure (generally less than 0.7 bar or 1 psi every 5 minutes) at the filtrate side indicates a passed test, while a significant decrease of the held pressure indicates a failed test.

Diffusive Air Flow Test

The diffusive air flow test uses the same concept as the air pressure-decay test, but is performed by monitoring the displaced liquid volume due to the leaking air from compromised fiber(s).

Bubble Test

Bubble testing can identify the fiber or seal location that is compromised in a membrane module. The test is typically performed after the compromised module is identified by a sonic sensor or any other monitoring method. After identifying the compromised fiber, it can be isolated from the module by adding an epoxy glue to its inlet, or by inserting pins of the same diameter as the fiber at the fiber inlet and outlet edges. The module can then be placed back into service.

Sonic Wave Sensing

Sonic sensor equipment consists of a sound wave sensor attached to a headphone. The headphones are manually placed at the top, middle, and bottom of the membrane module during the air-pressure decay test to detect any sound waves created by air bubbles leaking through a damaged fiber. The difference in sound between an intact and a compromised membrane may be identified by the Laboratory Testing Organization operators. Sonic sensing is only a qualitative tool for detecting loss of membrane fiber integrity, and therefore this test must be followed by a more quantitative method for evaluation of membrane integrity.

Experimental Objectives

The objective of this task is to demonstrate the methodology to be employed for monitoring membrane integrity at bench scale and to verify integrity of membrane modules.

Work Plan

The Laboratory Testing Organization shall clearly describe the most appropriate methods for monitoring of membrane integrity at bench scale. The techniques listed above are intended to serve as examples of direct methods for monitoring membrane integrity. These direct monitoring methods shall provide sensitive evaluation of membrane system integrity. It should be noted that pilot and/or full-scale methods of membrane integrity testing might have to be adapted for bench-scale applicability. Further, the bench-scale integrity monitoring does not replace integrity testing in the field. If the membrane module is shown to be compromised by integrity testing, it shall be discarded and another module shall be provided as a replacement.

Integrity testing shall be performed before and after challenge testing of each module. Since pressure decay tests (PDTs) are often used to measure membrane integrity, an example of adapting pilot integrity methodology to bench scale is described below.

PDTs are performed on each module before and after all challenge testing is completed for that module in order to assure the membrane module being evaluated is not compromised. To perform the test, the drain line is opened and the membrane fibers are emptied of feed solution by applying pressure from a nitrogen tank to the inner lumen of the fibers at the both ends of the modules (it is important, however, to keep the membrane fibers wetted). Nitrogen feed pressure is set at a pressure recommended by the manufacturer, after which the pressure feed line and drain line are closed. Then, the pressures at the inlet and outlet of the modules are monitored and recorded at intervals of not less than one per minute for a period recommended by the manufacturer. If the module is intact, only a small decrease in pressure is observed (usually less than 1 percent over a period of 15 minutes). If the membrane material or module is compromised, the

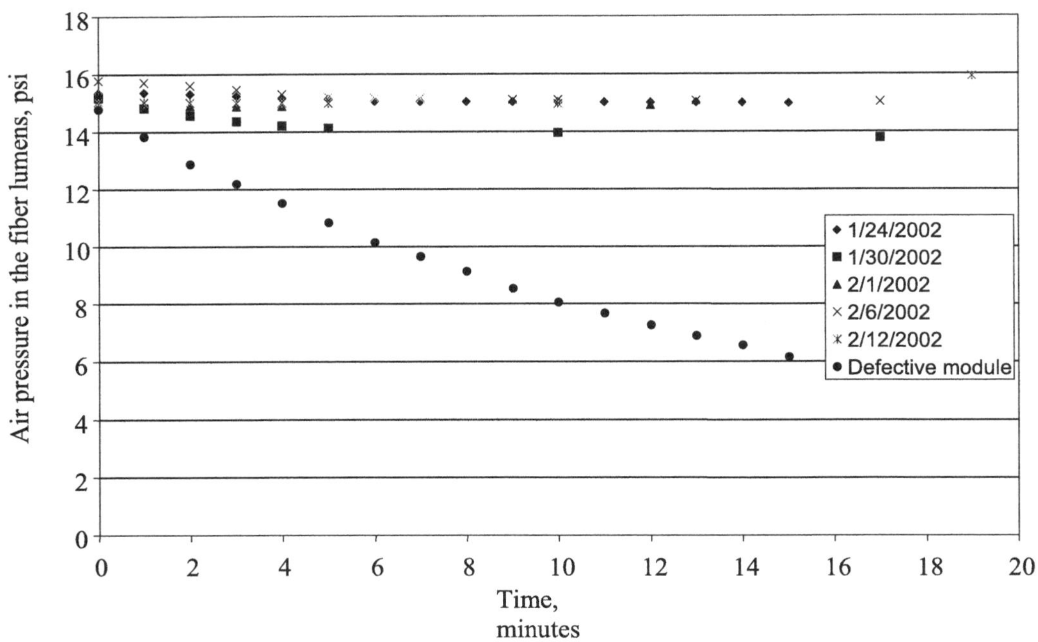

Figure 3.5 Pressure decay test performed on a 10-fiber, 30-inch long membrane module

PDT will show a substantial decrease of pressure over time. Figure 3.5 shows PDT data on both intact and compromised modules using a Low-Pressure Membrane Testing Unit.

Evaluation Criteria and Minimum Reporting Requirements

- Table of membrane integrity results before and after microbial challenge testing.
- Where appropriate for the selected integrity methodology, a temporal graph of integrity test results conducted before and after microbial challenge testing.

TASK D: CONDUCT MICROBIAL CHALLENGE EXPERIMENTS

Introduction

In this task, the effectiveness of membrane materials for microbial removal shall be evaluated by use of microbial challenge studies.

Experimental Objectives

The objective of this task is to characterize the low-pressure membranes in terms of microbial removal.

Table 3.2
Microorganisms for microbial challenges of low-pressure membranes*

Type of microorganism	Microorganism	Approximate size, μm
Protozoa	*Cryptosporidium parvum*	3-5
Bacteria	*Escherichia coli*	1-5
	Pseudomonas diminuta	0.6-1
Virus	MS2 bacteriophage	0.024
	PRD1 bacteriophage	0.070
	hepatitis A virus	0.025
	calicivirus	0.025

Source: Data from Joklik, et al., 1992; and Hunter 1997.
* The list of microorganisms in Table 3.2 is not a complete list; other microorganisms may be proposed for use as circumstances require.

Work Plan

The Laboratory Testing Organization shall conduct the microbial seeding studies as described in the sections below.

Organisms Employed for Bench-Scale Challenge Experiments

Table 3.2 presents the different microorganisms that may be employed for the bench-scale microbial challenge studies. One protozoan oocyst, two bacteria and four viruses can be used for the challenge studies. These organisms were chosen to provide a wide range in types and sizes of microorganisms in order to create a microbial removal profile for the low-pressure membrane being challenged. The list of microorganisms in Table 3.2 is not a complete list and other microorganisms may be proposed for use, as circumstances require.

It is recognized that, in many cases, it may not be possible to employ viable protozoan cysts and oocysts for challenge studies, depending upon the laboratory where the work is being performed. In such a case, the organisms shall be heat-fixed. Organism stocks received from appropriate suppliers shall be stored under refrigeration in the dark at 4°C or frozen (viruses only) with appropriate preservatives until use in the challenge studies. Methods for propagation and enumeration of the organisms listed in Table 3.2 are described or referenced in Appendix B.

Disinfection of Experimental System

Before performing microbial challenge experiments, the membrane module and tubing associated with the bench-scale Low-Pressure Membrane Testing Unit shall be disinfected using a free chlorine solution (or other appropriate biocidal agent), which is prepared in a pressurized feed tank. The concentration of the chlorine will be as recommended by the manufacturer. The pressure in the tank shall be set at 1 bar (15 psi) and the membrane unit shall be operated in a backwash mode.

The membrane module and associated tubing shall be backwashed for a minimum of three hydraulic cycles with the disinfecting solution. This module and associated tubing shall then be rinsed with a 3-molar excess sodium thiosulfate solution (or other appropriate chemical) to assure any residual chlorine is quenched. The membrane module shall then be rinsed in backwash mode at 1 bar (15 psi) for an additional three hydraulic cycles with a 0.1 mM phosphate buffer (pH 7.0).

Microbial Challenge Experiments

The microbial challenge experiments shall be conducted under the operating conditions in which the microorganisms would be most likely to penetrate the membrane. These conditions shall include the operational flux specified by the manufacturer for their membrane using a 0.1 mM phosphate buffer solution prepared in deionized water. All challenge testing shall be conducted as batch seeding tests under direct flow hydraulic conditions. The challenge testing shall be conducted for all organisms simultaneously, i.e., all organisms shall be seeded into the feedwater prior to conducting the testing. Only organisms that do not cross react should be employed. Further, tests shall be conducted to detect and quantify any microbial adsorption that occurs onto the non-membrane parts of the Low-Pressure Membrane Testing Unit. This may be accomplished by conducting a microbial challenge control test with a module that does not contain potted membranes.

For each module tested, four samples shall be collected: two discrete seeded feed tank samples (at the beginning and at end of the each test) and two discrete filtrate samples (samples may be collected sequentially, one right after the other). Thus, for the five modules evaluated, a total of 20 samples shall be collected.

Feedwater to which microorganisms shall be added shall consist of a 0.1 M phosphate buffer (pH 7.0) prepared from sterile, deionized laboratory water. To check the quality of the water, measurements of pH, turbidity, particle counts, and conductivity shall be made and recorded before the seeding of any organisms. Additionally, the TOC concentration shall be less than 0.2 mg/L. Feedwater turbidity shall not exceed 0.1 NTU. Total particle counts in the 2 - 50 μm size range shall not exceed 25 per mL. Methods for these analyses are described in Appendix B.

The feed suspension of microorganisms shall be prepared by adding the concentrated stock suspension(s) of organisms into the feedwater reservoir. For organisms that are propagated at very high titers (for MS2 and PRD1, the initial stock densities are approximately 10^{11} to 10^{12} pfu/mL), one or more dilutions shall be made before adding the organisms to the feedwater tank. This tank shall be completely mixed during preparation of the seeded feedwater. Sufficient volume of stock suspension shall be created to sustain membrane operation for a minimum of 8 HRTs per membrane module per microbial challenge test. After the addition of challenge organism(s) to the feedwater tank and before the initiation of filtration, one discrete sample shall be collected from the feedwater tank to establish the initial titer of the microorganisms. Each sample shall consist of collecting 35 mL of filtrate in a sterile, 50-mL polypropylene centrifuge tube. Polypropylene is employed to avoid adsorption of the microorganisms onto the walls of tube. For the protozoan challenge tests, the final seeding concentration in the feedwater tank shall be high enough to provide a removal sensitivity limit of at least 4 logs. For the bacterial and viral challenge tests, the final seeding concentration in the feedwater tank shall also be high enough to provide a removal sensitivity limit of at least 5 logs. Detection limits for microorganisms used in challenge testing shall be noted in the QA/QC plan.

After collecting feedwater samples, the membrane shall be operated under the hydraulic conditions (pressure and filtrate flux) established under Task A. The feed suspension of microorganisms shall be filtered under these conditions for a total of five HRTs in order to achieve steady state. At the end of this period, two discrete, consecutive samples shall be collected from the feed tank. Each sample shall consist of collecting 35 mL of filtrate in a sterile, 50-mL polypropylene centrifuge tube. At the conclusion of testing each module, a second sample shall be collected from the feedwater tank. All samples shall be stored at 4°C immediately after collection. The $\log_{10}$ mean values of the "before" and "after" feedwater tank microbial densities shall be used when calculating microbial removal.

Before beginning the testing of the next previously-conditioned membrane module, the module and tubing shall be backwashed with 50 mg/L of free chorine (or other appropriate biocidal agent) for five hydraulic cycles. The membrane shall then be rinsed with a 3-molar excess sodium thiosulfate solution (or other appropriate chemical) buffer and phosphate buffer as described above. After this procedure, the next test shall be initiated. At the end of each day, the microbial samples shall be shipped via overnight express for enumeration, if not enumerated on site.

Operational Data Collection

Operational data are collected under Task A.

Evaluation Criteria and Minimum Reporting Requirements

- Table of water quality data: pH, turbidity, particle counts, and conductivity
- Table of feedwater and filtrate levels for all organisms
- Bar chart of removal ($\log_{10}$) of microorganisms

TASK E: EXECUTE DATA HANDLING PROTOCOL

Introduction

The data management system used in the bench-scale membrane characterization shall involve the use of computer spreadsheets and manual recording of operational parameters for the Low-Pressure Membrane Test Unit.

Experimental Objectives

The objective of this task is to establish a structure for the recording and transmission of laboratory testing data.

Work Plan

The following protocol has been developed for data handling and data verification by the Laboratory Testing Organization. Specific parcels of the computer databases for operational and water quality parameters shall be entered by manual importation into Excel (or similar spreadsheet software). Backup of the computer databases to diskette, compact disk, magnetic tape or other archival format shall be performed at the end of each day.

Measurements shall be recorded on specially prepared data log sheets as appropriate. A laboratory notebook shall be used to record all data, calculations and other pertinent information not included in the data log sheets. The laboratory notebook shall provide carbon copies of each page. The original notebooks shall be stored in the laboratory.

The database for the project shall be set up in the form of custom-designed spreadsheets. The spreadsheets shall be capable of storing and manipulating each monitored water quality and operational parameter from each task, each sampling location, and each sampling time. All data from the laboratory notebooks and data log sheets shall be entered into the appropriate spreadsheet. All recorded calculations shall also be checked at this time. Following data entry, the spreadsheet shall be printed out and the hard copy shall be checked against the handwritten data sheet. Any corrections shall be noted on the hard copies and corrected on the screen, and then a corrected version of the spreadsheet shall be printed out. Each step of the verification process shall be initialed by the bench testing operator, technician or engineer performing the entry or verification step.

The testing of each membrane module shall be assigned a run number that will then be tied to the data from that experiment through each step of data entry and analysis. As samples are collected, the data shall be tracked by use of the same system of run numbers. Data from the outside laboratories, if any, shall be received and reviewed by the laboratory staff conducting the studies. These data shall be entered into the data spreadsheets, corrected, and verified in the same manner as the field data.

CHAPTER 4
RESULTS

BENCH-SCALE STUDIES – HOLLOW FIBER MEMBRANES

Removal of MS2 Bacteriophage by Selected Low-Pressure Membranes

Initial studies focused on the removal of MS2 bacteriophage by a wide variety of membranes. Table 4.1 provides information about the low-pressure membranes employed. In total, seven sets of membranes composed of PVDF or a modified polysulfone (mPS) were evaluated. All membranes had a feed-flow configuration of outside-in and ranged in manufacturer-reported MWCO or pore size from 10 kD to 0.2 μm. Two independent sets of mPS membranes with the same MWCO were part of this group (mPS 100 kD and mPSa 100 kD). All membranes were tested using the bench-scale apparatus described in Chapter 3, except for the PVDF 0.035 μm membrane, which was operated on a small-scale pilot system provided by the manufacturer. For this reason, a larger membrane area (1.0 m^2) was employed for this membrane as compared to the other membranes (0.1 m^2). As noted in Table 4.1, all membranes operated at different clean water specific fluxes, which were consistent with the various manufacturer-reported MWCOs or pore sizes.

Figure 4.1 shows the removal of MS2 bacteriophage by the various membranes. Error bars represent one standard deviation from the mean. The lowest bacteriophage removal (<1.0 log) was observed for the PVDF 0.1 μm and the mPS 0.2 μm membranes. Despite the low bacteriophage removal, the null hypothesis that they were not statistically different than zero removal was rejected at $p = 0.01$. The mPS 100 kD and 300 kD membranes removed the bacteriophage to the sensitivity limits of the assay. However, it is important to note that the mPS 10 kD membrane was less effective in removing MS2 bacteriophage than the "looser" membranes with the highest manufacturer-reported MWCOs; this observation is discussed in more detail below.

Impact of Microorganism Type

Use of MS2 and PRD1 Bacteriophage as Surrogates for Enteric Viruses

Although enteric viruses were evaluated as part of this study, they are time-consuming to culture and assay. As a result, experiments were conducted to evaluate if surrogate organisms for membrane challenge studies could be employed. The traditional organism employed for such studies is MS2 bacteriophage (USEPA 2002). However, there are little data in the literature describing parallel microbial challenge studies using this bacteriophage and enteric viruses. In this study, microbial challenge experiments were conducted by exposing selected membranes (PVDF 0.1 μm, mPS 0.2 μm, mPS 10 kD, mPS 100 kD and mPS 300 kD) to a mixed suspension of MS2 and PRD1 bacteriophage and various enteric viruses under conditions varying in trans-membrane flux, ionic strength, pH and humic and fulvic acid concentrations. Adding all of the organisms concurrently (as opposed to sequentially) eliminated any experiment-to-experiment variation.

Figure 4.2 shows the correlation between MS2 bacteriophage and all of the enteric viruses studied (PV1, HAV and FCV). A line of perfect correlation is presented for comparison purposes.

Table 4.1
Membranes employed in bench study

Module designation	mPs 10 kD	mPS 100 kD	mPSa 100 kD	mPs 300 kD	PVDF 0.035 μm	PVDF 0.1 μm	mPS 0.2 μm
Manufacturer	Polymem	Polymem	Polymem	Polymem	Zenon	Memcor	Polymem
Reported MWCO (kD)*	10	100	100	300	n/a	n/a	n/a
Nominal pore size (μm)*	n/a	n/a	n/a	n/a	0.035 μm	0.1	0.2
Material	Modified PS	Modified PS	Modified PS	Modified PS	PVDF	PVDF	Modified PS
Geometry	HF	HF	HF	HF	HF	HF	HF
Flow configuration	O/I	O/I	O/I	O/I	O/I	O/I	O/I
Membrane surface area (m^2)	0.1	0.1	0.1	0.1	1	0.1	0.1
External diameter (mm)	0.72	0.72	0.72	0.72	0.7	0.65	0.72
Internal diameter (mm)	0.38	0.38	0.38	0.38	0.4	0.39	0.35
Max. operating TMP† (bar)	2.4	2.4	2.4	2.4	0.7	1	2.4
Clean Water J_s ($L/m^2/hr/bar$)	60	200	250	160	200	370	400

* As reported by the manufacturer
† As recommended by the manufacturer
n/a: not applicable
HF: hollow fiber
O/I: outside/in

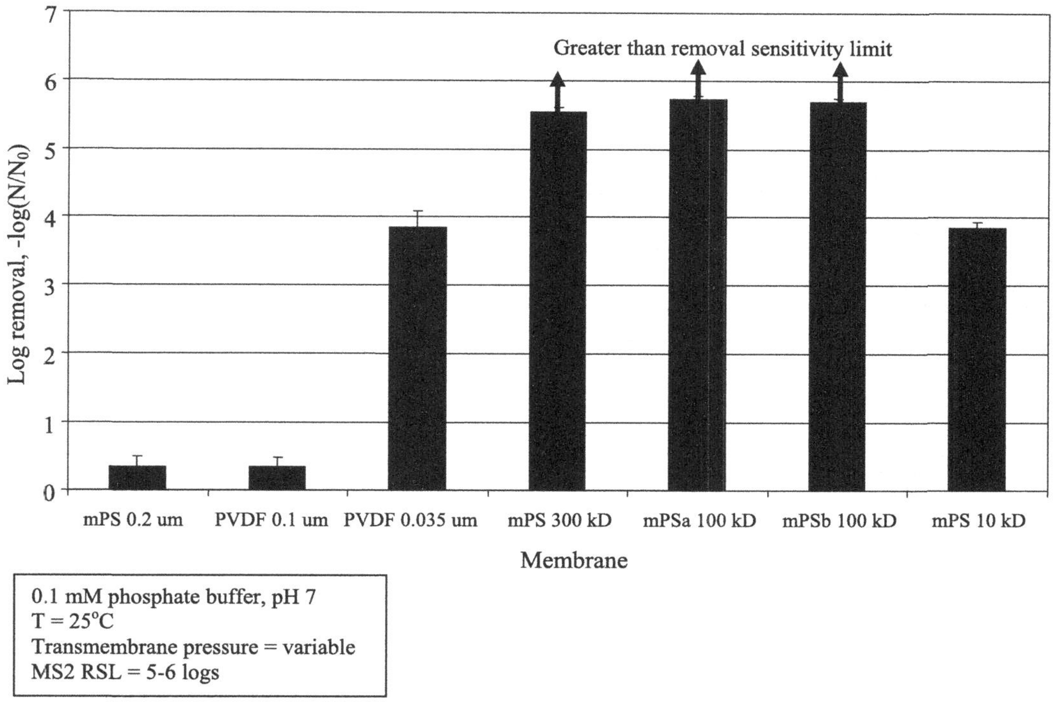

Figure 4.1 Removal of MS2 bacteriophage by selected low-pressure membranes

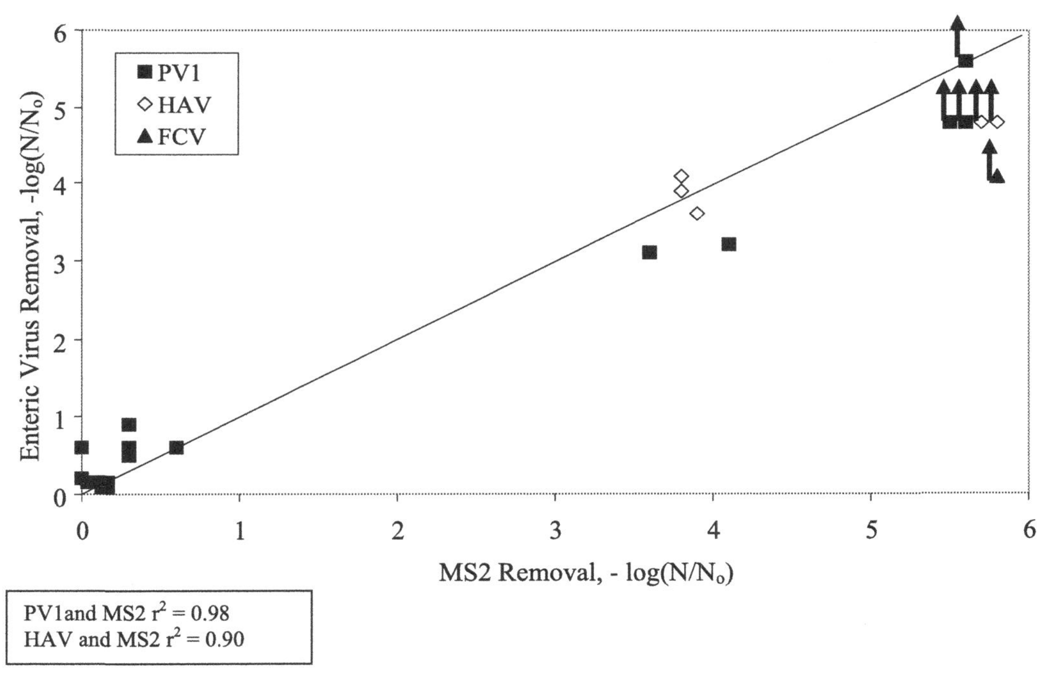

Figure 4.2 Correlation of MS2 bacteriophage and enteric virus removal by various low pressure membranes

Table 4.2
Matched-pair analysis showing mean differences and confidence intervals between removal of selected bacteriophage and enteric viruses

Surrogate organism	Parameter	Poliovirus 1	Hepatitis A virus	All enteric viruses
MS2	Mean difference without RSL data	-0.06	-0.10	-0.07
bacteriophage	Confidence interval around mean without RSL data at P=0.05	0.16	0.21	0.15
	Mean difference with RSL data	-0.05	-0.04	-0.04
	Confidence interval around mean with RSL data at P=0.05	0.12	0.11	0.09
PRD1	Mean difference without RSL data	0.51	0.66	0.56
bacteriophage	Confidence interval around mean without RSL data at P=0.05	0.40	0.24	0.28
	Mean difference with RSL data	nd	0.33	0.42
	Confidence interval around mean with RSL data at P=0.05	nd	0.24	0.24

RSL = Removal sensitivity limit of bacteriophage or enteric virus assay
nd = no data

The correlations of MS2 bacteriophage with PV1 and HAV were both 0.90 or greater. It should be noted that all of the data points at 5 logs of removal or greater were above the sensitivity limit of the assay for each organism. The variation in sensitivity limits is the reason for data points being below the line of perfect correlation.

The data were also analyzed using a matched-pair analysis since, as noted above, all of the organisms were exposed concurrently in each individual challenge experiment. The differences in removal of individual bacteriophage and enteric viruses were calculated for each experiment. The mean difference and confidence interval for this population mean (at P = 0.05) were then determined for each matched surrogate organism and virus. From these data, it was determined if the mean difference was statistically greater than zero. The analysis was conducted with and without sensitivity limit values included. Table 4.2 shows that the mean differences in matched MS2 bacteriophage and enteric virus pairs (shown as individual virus and aggregated as all enteric viruses) were not statistically different than zero. Including the sensitivity limit values for the bacteriophage and virus data in the analysis did not change the outcome. The results obtained suggest that MS2 bacteriophage served as a good surrogate for the enteric viruses under the conditions studied.

In comparison, mean differences in matched PRD1 and enteric virus pairs (also shown as individual virus and aggregated as all enteric viruses) ranged from 0.51 to 0.66 log. These

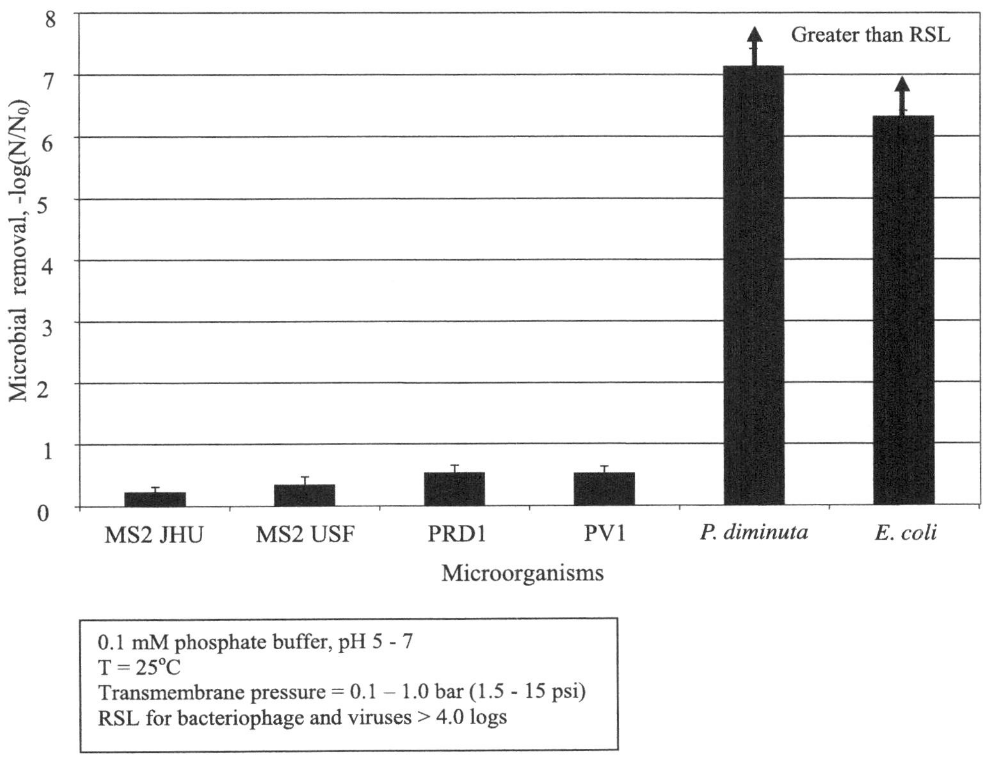

Figure 4.3 Removal of selected microorganisms by PVDF 0.1 μm membrane

differences were statistically different than zero at P = 0.05. That PRD1 bacteriophage was removed to a greater degree than the enteric viruses suggests that this organism would not serve as a good surrogate. However, PRD1 bacteriophage could be used in conjunction with MS2 bacteriophage in microbial challenge experiments to better define the range of microorganisms that could be removed by a particular low pressure membrane.

Impact of Microorganism Type on Microbial Removal

This study focused on two different types of microorganisms: virus and bacteria. Protozoa were not part of the scope of work; however, since protozoa are as large (1-15 μm) or larger than most bacteria (0.2-10 μm), it is probable that any protozoa removal based on size exclusion would be similar or greater than that of bacteria. Figure 4.3 shows the differences in removal of virus and bacteria by a selected membrane (PVDF 0.1 μm). Controls on MS2 bacteriophage assays were conducted by analyzing split samples at the University of South Florida (USF) and the Johns Hopkins University (JHU) for selected experiments. Results of the split samples are shown in the figure. The paired control MS2 bacteriophage assays conducted at JHU and USF showed that the hypothesis that there was no statistical difference between the two was rejected (P = 0.24). These results provide confidence in the validity of the data reported.

Figure 4.3 shows that *E. coli* and *P. diminuta* were exposed to the membrane at concentrations greater than 10^6 cfu/mL. In contrast to the observed virus removal, the bacteria were removed to the sensitivity limits of their respective assays in all experiments. Less than 1 log of

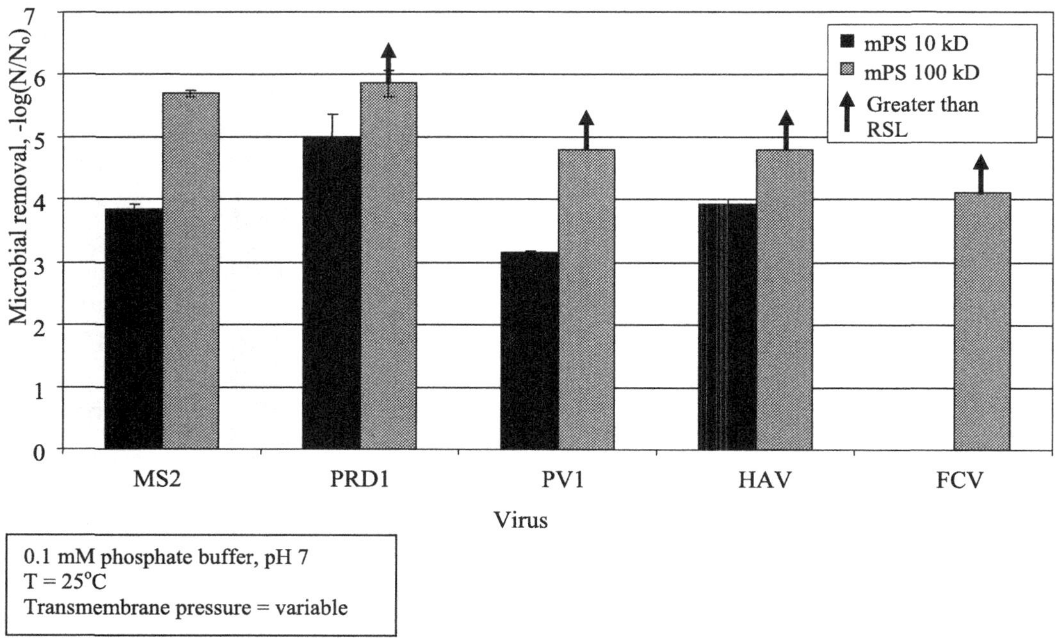

Figure 4.4 Removal of selected enteric viruses and bacteriophage by two low-pressure membranes (mPS 10 kD and mPS 100 kD)

bacteriophage and virus removal by the PVDF 0.1 μm membrane was observed. However, the data suggest that PRD1 bacteriophage was removed to a greater extent than MS2 bacteriophage and poliovirus 1 (PV1), probably due to it larger size (70 μm, 24 μm and 25 μm for PRD1 bacteriophage, MS2 bacteriophage and PV1, respectively). An analysis of variance showed that that there was a statistical difference between the removal of three viruses at $P = 3.2 \times 10^{-4}$. The difference in removal between MS2 and PRD1 bacteriophage was significant at $P = 9.1 \times 10^{-5}$. There was no statistical difference between the removal of PRD1 bacteriophage and PV1 or between MS2 bacteriophage and PV1.

Greater focus on the removal of specific viruses was employed using two mPS membranes (mPS 10 kD and mPS 100 kD). Figure 4.4 shows the removal of MS2, PRD1, PV1, hepatitis A virus (HAV) and feline calicivirus (FCV) at pH 7 and 25°C. The data show that the mPS 100 kD removed both the enteric viruses and the bacteriophage to their respective removal sensitivity limits (PRD1 bacteriophage, PV1, HAV and FCV) or near these limits (MS2 bacteriophage). All were removed by greater than 4.0 logs. In comparison, between 3.0 and 4.0 logs of MS2, PV1 and HAV were removed by the mPS 10 kD membrane. There was no significant difference in removal of MS2 bacteriophage and HAV ($P = 0.41$). However, significant differences were observed between the removal of MS2 and PRD1 bacteriophage ($P = 3.0 \times 10^{-4}$) and MS2 bacteriophage and PV1 ($P = 2.8 \times 10^{-4}$). Once again it is important to note that the mPS 10 kD was less effective in removing MS2 bacteriophage than the "looser" membranes with high manufacturer-reported MWCOs.

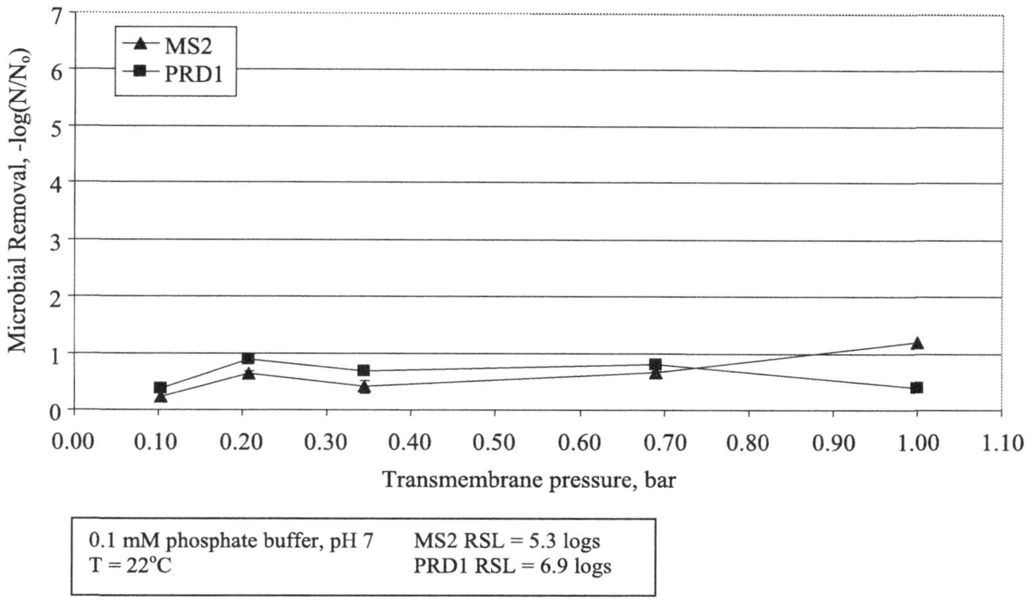

Figure 4.5 Impact of transmembrane pressure on removal of MS2 and PRD1 bacteriophage by PVDF 0.1 μm membrane

Impact of Transmembrane Flux

The impact of transmembrane pressure and flux on virus removal was evaluated in a series of experiments over the pressure range of 0.1 to 2.1 bar (1.5 psi to 30 psi). In order to assess the impact of these parameters on a wide range of filtrate virus concentrations, membranes expected to remove various levels of virus were employed: one was expected to remove less than 1 log (PVDF 0.1 μm) and one between 3 and 5 logs (mPS 10 kD); a third (mPS 100 kD) was expected to remove the viruses to the sensitivity limits of the respective assays. All experiments were conducted at pH 7 and approximately 25°C. An example of results for the PVDF 0.1 μm membrane is shown in Figure 4.5. Less than 1 log of removal was observed at all transmembrane pressures evaluated for both MS2 and PRD1 bacteriophage. A Student's t test was conducted on the slopes of the lines associated with the regression analysis to test the null hypothesis that there was no impact of pressure. The hypothesis was not rejected at $P < 0.01$. Similar experiments were conducted for the mPS 10 kD (Figure 4.6) and mPS 100 kD (Figure 4.7) using a variety of viruses. The data show that specific flux had little impact on the removal of the target organisms.

Impact of pH

As shown in previously, the PVDF 0.1 μm membrane removed less than 1 log of MS2 and PRD1 bacteriophage and PV1. Further experiments were conducted to evaluate the impact of pH on the removal of these organisms. Three different pH levels were employed: 5, 7 and 9. All experiments were conducted at specific fluxes ranging from 270 to 400 L/m^2/hr/bar (11 to 16 gfd/psi). The results are shown in Figure 4.8. For a pH ranging from 5 to 9, there was little impact on removal of any of the bacteriophage or virus. An analysis of variance was conducted on the mean

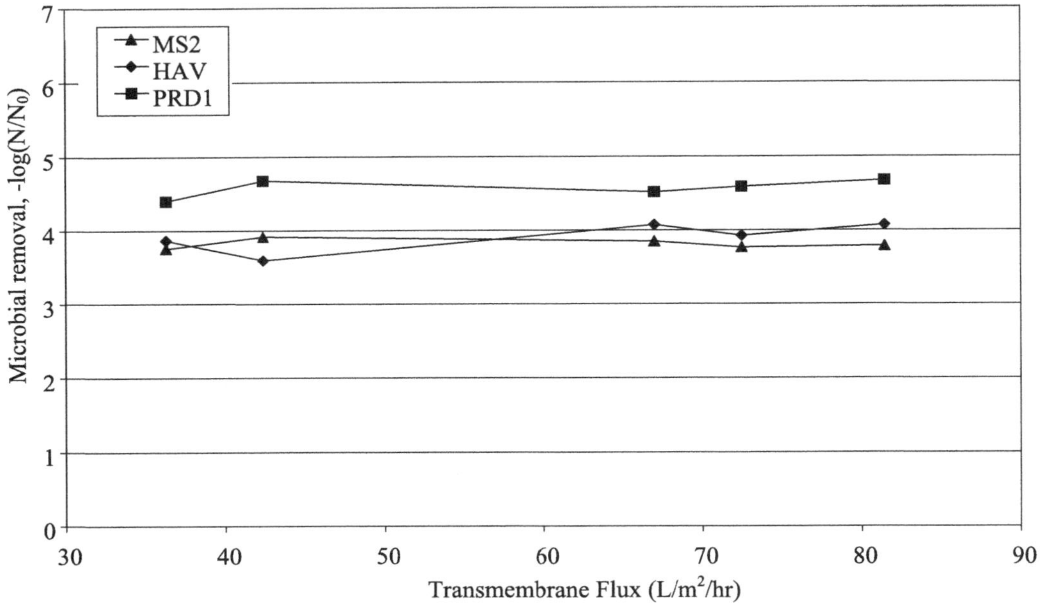

0.1 mM phosphate buffer, pH = 7.0
T = 26°C
MS2 RSL = 5.7 logs
PRD1 RSL = 5.8 logs
HA V RSL = 4.8 logs

Figure 4.6 Impact of transmembrane flux on the removal of MS2 and PRD1 bacteriophage and hepatitis A virus by mPS 10 kD membrane

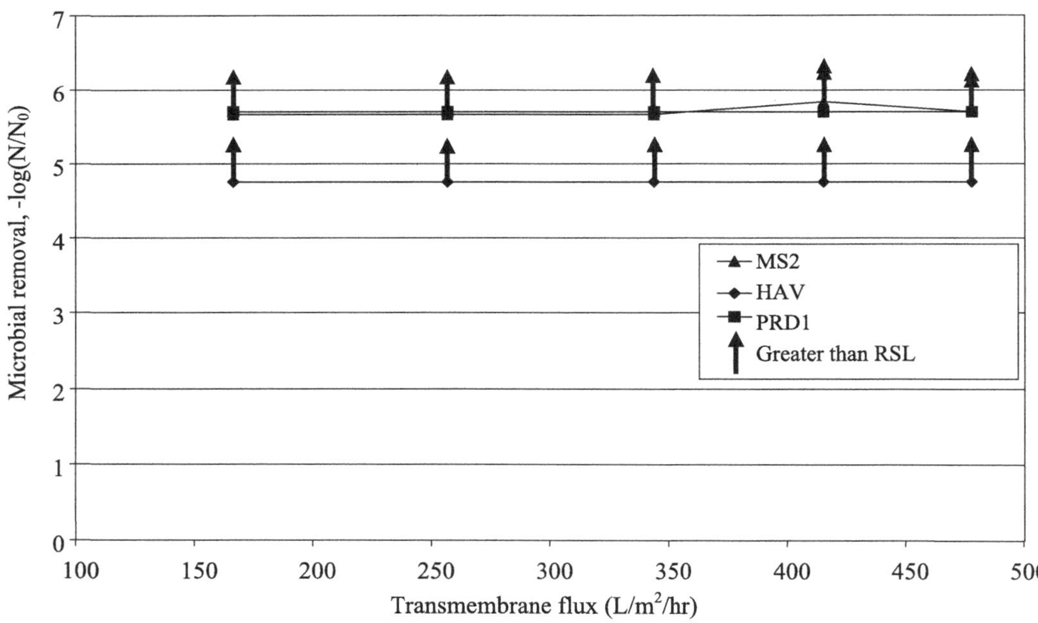

0.1 mM phosphate buffer, pH = 7.0
T = 26°C
MS2 RSL = 5.7 logs
PRD1 RSL = 5.6 logs
HA V RSL = 4.8 logs

Figure 4.7 Impact of transmembrane flux on the removal of MS2 and PRD1 bacteriophage and hepatitis A virus by mPS 100 kD membrane

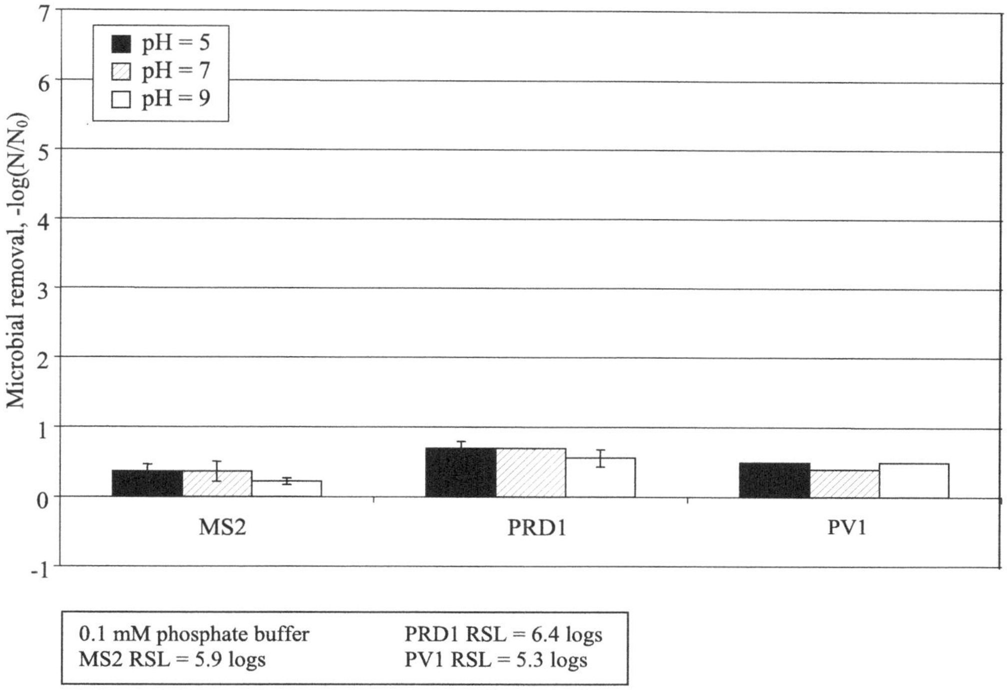

Figure 4.8 Impact of pH on the removal of MS2 and PRD1 bacteriophage and poliovirus 1 by PVDF 0.1 μm membrane

rejection values for MS2 and PRD1 bacteriophage. There was no statistical difference in the mean bacteriophage removal due to pH for MS2 ($P = 0.57$) and for PRD1 ($P = 0.65$).

Impact of Phosphate Buffer and NaCl Concentration

Since phosphate buffer was used in all clean water experiments, its impact on virus removal was examined for the PVDF 0.1 μm membrane. Phosphate buffer concentrations of 0.1 mM, 1 mM and 10 mM were employed. The corresponding conductivities are shown in Figure 4.9. The data show that the first two concentrations had little impact on virus removal. However, at the highest concentration evaluated, approximately 2.0 logs of MS2 and PRD1 bacteriophage removal and 4.3 logs of PV1 removal were observed. The results suggest that phosphate buffer did not have an impact on virus removal at concentrations used in this study (0.1 mM).

The impact of salt concentration on the removal of microorganisms was also evaluated using sodium chloride (NaCl). Conductivity was modified by increasing the NaCl concentration in the feed solution.

The results are shown in Figure 4.10. The data show increasing bacteriophage and virus removal with increasing salt concentration; however, very high concentrations (approximately 10,000 mg/L) were necessary to attain greater than 3 logs of MS2 and PRD1 bacteriophage removal.

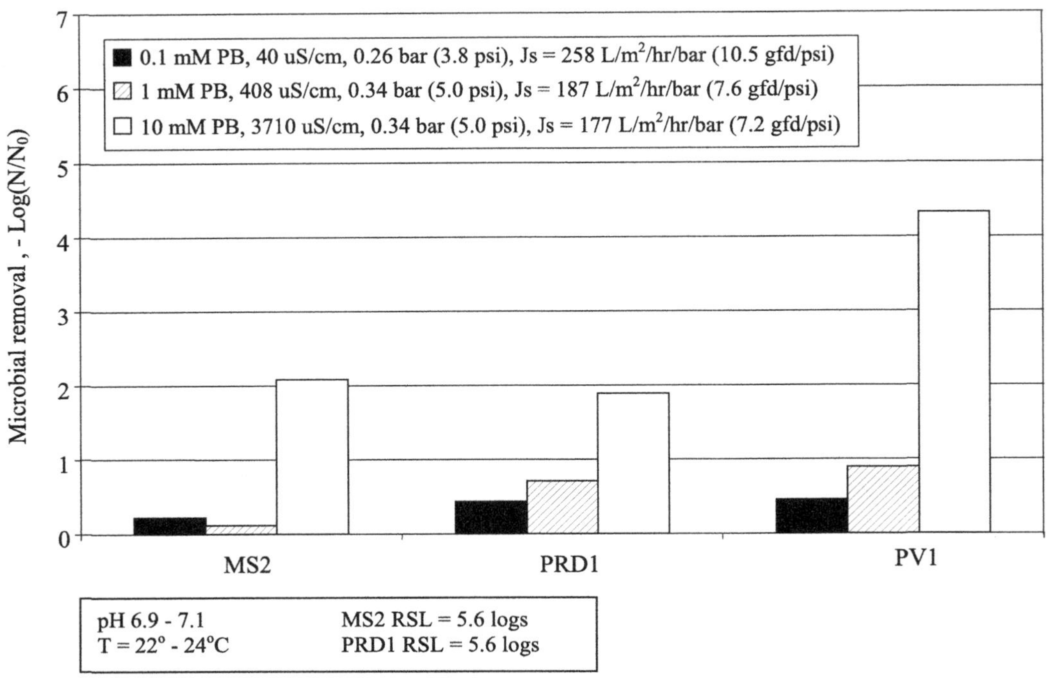

Figure 4.9 Impact of phosphate buffer on the removal of MS2 and PRD1 bacteriophage by PVDF 0.1 μm membrane

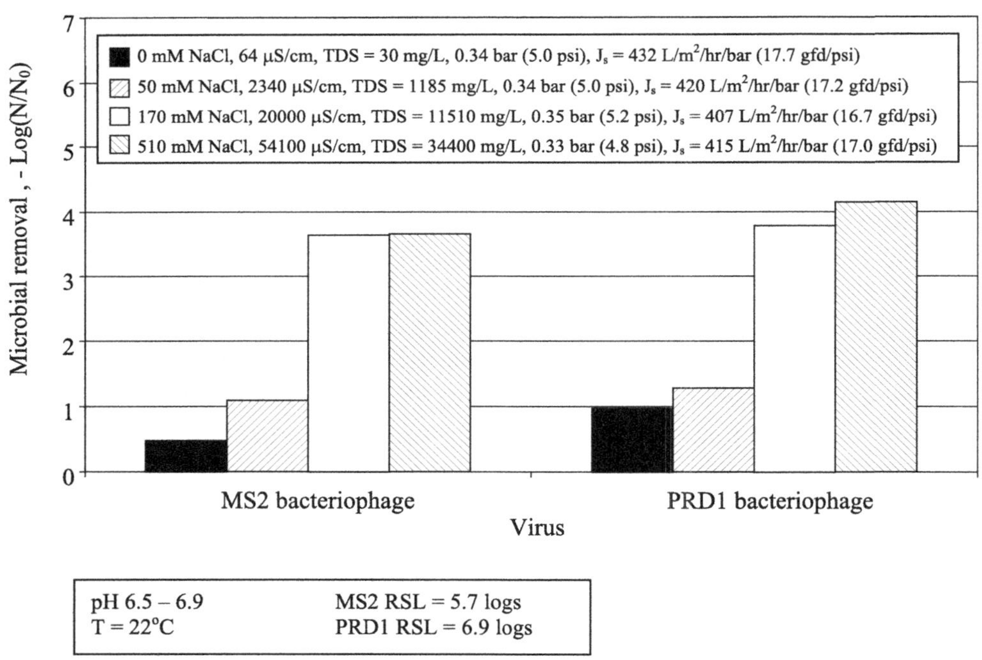

Figure 4.10 Impact of NaCl concentration on the removal of MS2 and PRD1 bacteriophage by PVDF 0.1 μm membrane

Table 4.3
Characteristics of a stock bentonite suspension employed in solids deposition experiments

Parameter (units)	Value
pH	6.9
T (°C)	27
Conductivity (μS/cm)	25
Suspended solids (g/L)	0.8
Turbidity (NTU)	88
Mean size (μm)	1.4
Particle count (#1 μm/mL)	1.68×10^7

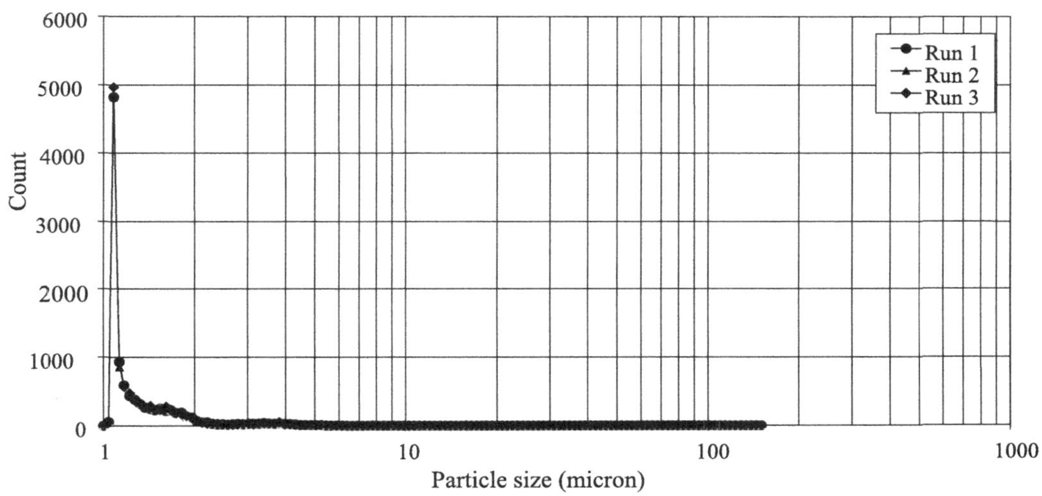

Figure 4.11 Distribution of bentonite particles by size in diluted stock suspension used for solids deposition experiments

Impact of Deposition Layer

Impact of a Bentonite Pre-coating on the Membrane

Experiments were conducted to evaluate the impact on bacteriophage and virus removal of a solids deposition layer on the membrane surface. A bentonite suspension was employed for the pre-coating the membrane. In these experiments, bentonite was added to a target mass loading on the membrane surface, followed by challenging the membrane with a bacteriophage and virus feed suspension. The characteristics of a stock bentonite preparation are shown in Table 4.3. A suspension of 0.8 g/L of bentonite having a conductivity of 25 μS/cm yielded a turbidity of 88 NTU and a particle concentration of 1.7×10^7 particles > 1 μm/mL. Figure 4.11 shows the particle size distribution of a diluted suspension of bentonite for three replicate experiments. The data

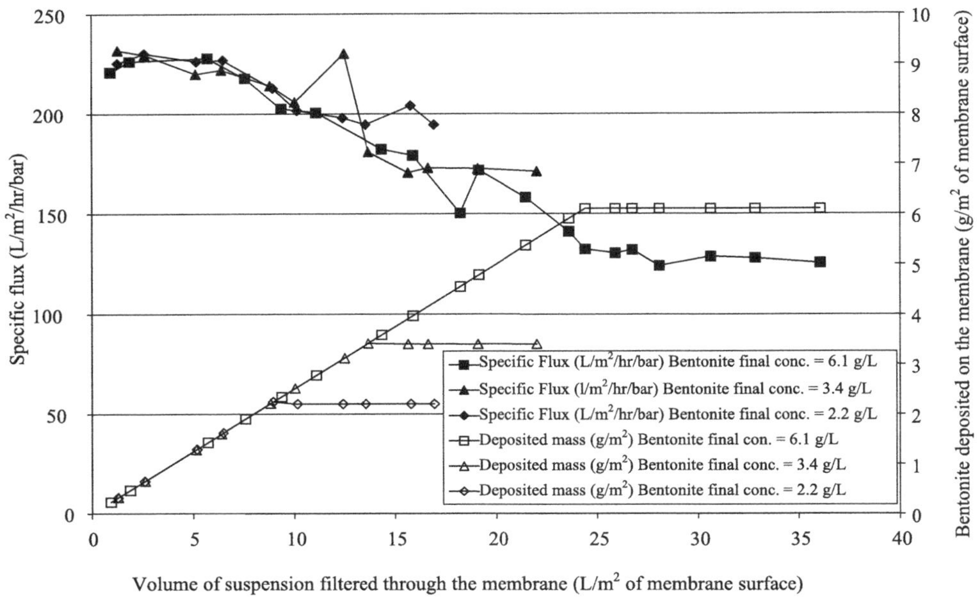

Figure 4.12 Impact of bentonite mass loading (250 mg/L suspension) on specific flux of PVDF 0.1 μm membrane at 1 bar (15 psi) and pH 7

show that the mean particle size was 1.1 μm with over 90 percent of the particles between 1 and 2 μm. Thus, any effects of a bentonite deposition layer was primarily attributable to this particle size range.

Pre-coating the membrane with bentonite impacted the operational aspects of the membrane. Figure 4.12 shows the impact on specific flux of a 250-mg/L suspension of bentonite loaded onto the membrane surface at 1 bar (15 psi) and pH 7. As expected, as the solids loading increased, the specific flux decreased. It is important to note that as soon as the loading of bentonite onto the membrane surface was stopped, the specific flux remained constant. For example, when 23 L of the bentonite solution per m^2 of membrane area was applied to the membrane (6 g/m^2), the specific flux declined to approximately 150 $L/m^2/hr/bar$ (5 gfd/psi). However, when the bentonite feed suspension was stopped, specific flux remained constant. These data suggest that the continual operation of the membrane at a constant pressure does not promote further compaction of the deposition layer. A summary of the deposition layer experiments on specific flux is shown in Figure 4.13 at 0.3 bar (5 psi) and pH 6.4 to 7.1; a specific flux decline of over 85 percent was observed at high solids loading rates (15 to 20 g/m^2 of bentonite).

Two types of bacteriophage and virus challenge experiments were conducted. The first evaluated the impact of flux at a set mass loading of bentonite (16 g/m^2). The second evaluated the impact of different bentonite mass loadings on microbial removal. Figure 4.14 shows the impact of flux on the removal of MS2 bacteriophage and poliovirus at a constant bentonite loading and varying pressures. The data show that increasing the transmembrane pressure (i.e., increasing the specific flux) did not increase microbial removal. For MS2 bacteriophage, between 2.5 and 3.0 logs of removal were achieved irrespective of specific flux. Consequently, it appears that compacting the cake layer by applying pressures above 0.3 bar (5 psi) did not increase microbial removal. Similar results were observed for poliovirus; however, since the removal sensitivity

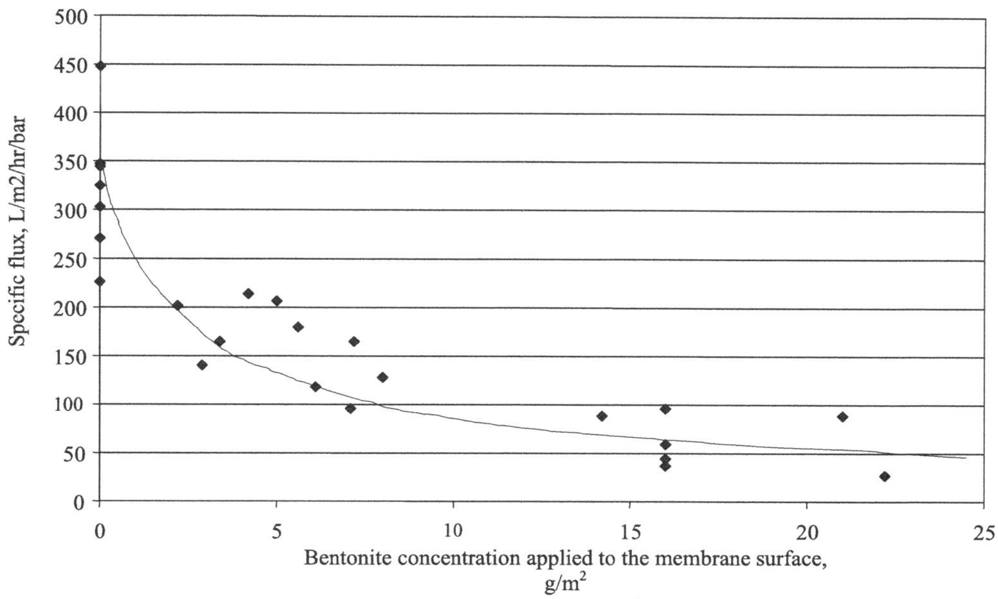

Figure 4.13 Impact of bentonite mass loading on specific flux of PVDF 0.1 μm membrane specific flux at pH 7 and 25°C

limits were low for this virus, it was difficult to clearly assess what the upper level of removal may have been.

Figure 4.15 presents the impact of bentonite mass loading on removal of bacteriophage and virus. The data show that in the range of 0.0 to 3.8 g/m^2 of bentonite deposited on the membrane surface, the removal of MS2 bacteriophage increased from 0.2 to 2.7 log and the removal of poliovirus increased from 0.6 to 4.3 logs. As expected, the bentonite formed an additional barrier to virus passage, the extent of which was semi-quantifiable.

Impact of Manufacturer-Reported Molecular Weight Cut-Off or Pore Size

One important aspect of this study was to evaluate the validity of using manufacturer-reported MWCO or pore size for estimating microbial removal. The impact of these parameters on bacteriophage removal was evaluate using five different hollow fiber membranes (mPS 10 kD, mPS 100 kD, mPSa 100 kD, mPS 300 kD, mPS 0.2 μm). These membranes were chosen because the only variable was the reported MWCO or pore size. Each membrane was composed of similar material (modified polysulfone) and provided by the same manufacturer, who potted the membrane according the specifications outlined in Section 3. All membranes, which were pressure tested before and after each experiment, were integral. Membrane MWCOs reported by the manufacturer ranged from 10 kD to 300 kD. Additionally, one membrane had a 0.2 μm reported nominal pore size. Other characteristics are shown in Table 4.1.

A set of two membranes for each MWCO or pore size was challenged and the results averaged. For the mPS 100 kD membrane, two similar sets of two membranes were provided (referred to as "mPS 100 kD" and "mPSa 100 kD"). The membranes were exposed to 5.5 to 6.2 logs of the microorganisms and operated at a transmembrane pressure of 1.5 to 1.8 bar (22 to 26 psi).

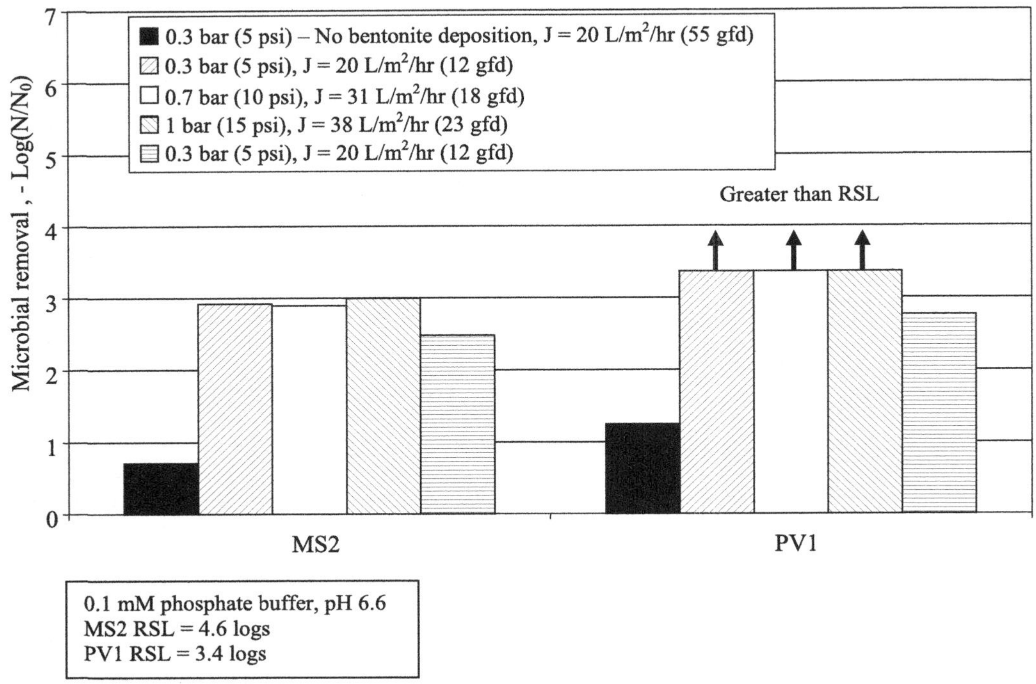

Figure 4.14 Impact of transmembrane pressure on the removal of MS2 bacteriophage and poliovirus 1 by PVDF 0.1 μm membrane with 16 g/m² of bentonite loading

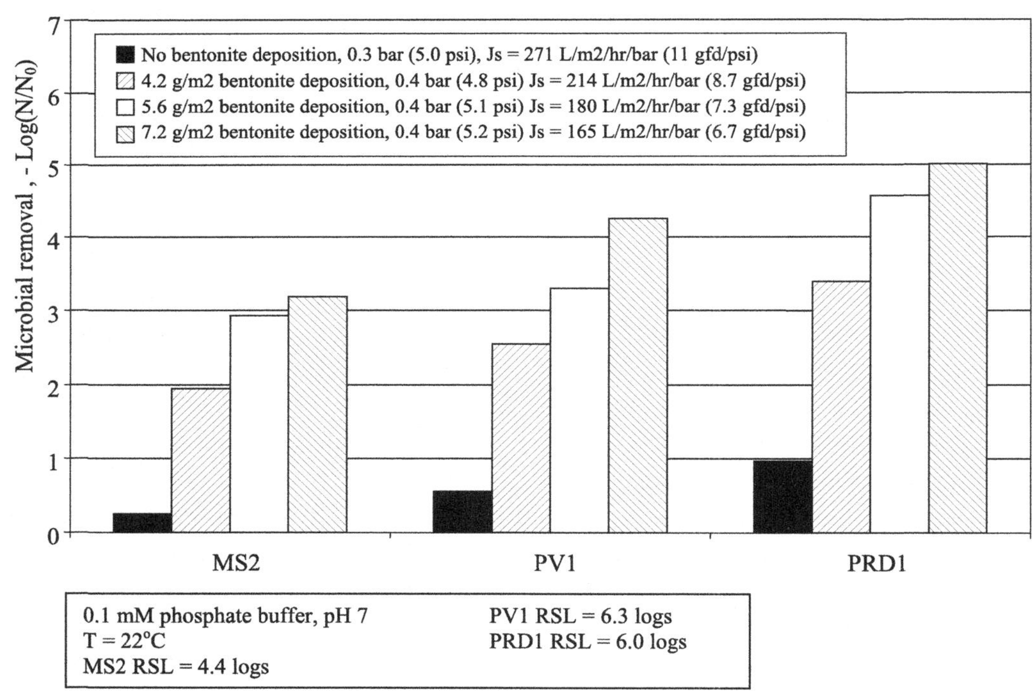

Figure 4.15 Impact of bentonite mass loading on the removal of MS2 and PRD1 bacteriophage and poliovirus 1 by PVDF 0.1 μm membrane

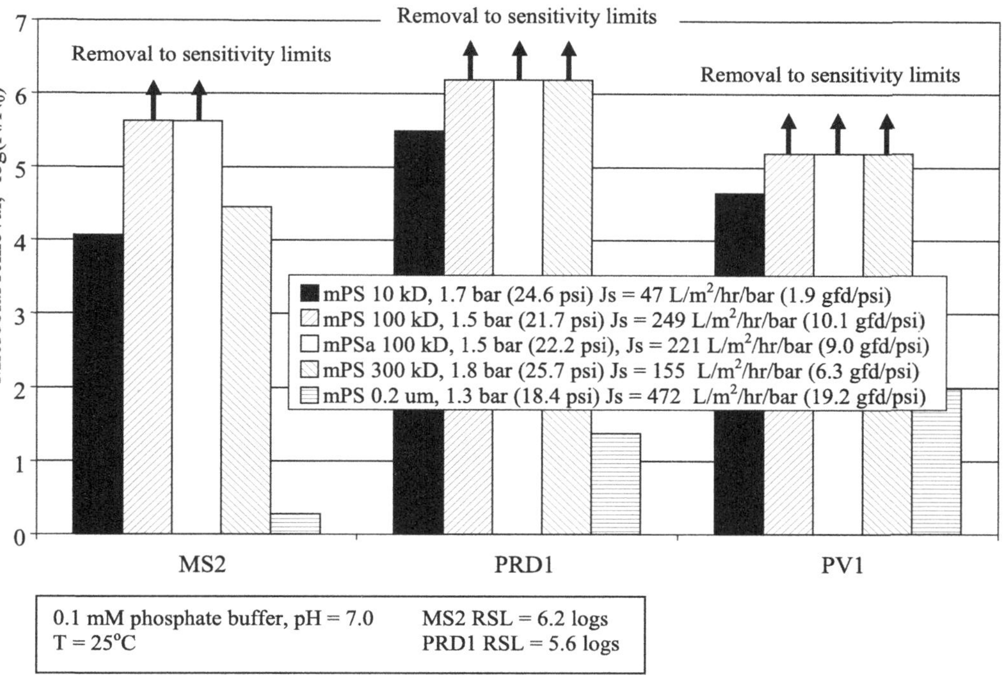

Figure 4.16 Removal of MS2 and PRD1 bacteriophage and poliovirus 1 by various modified polysulfone membranes

Figure 4.16 shows the data from these experiments. As expected, the lowest MWCO membrane had the lowest specific flux (47 L/m^2/hr/bar, 1.9 gfd/psi). Conversely, the most porous membrane (0.2 μm) had the highest specific flux (472 L/m^2/hr/bar, 19.2 gfd/psi). However, the removal of MS2 bacteriophage did not correspond to either increasing MWCO or increasing specific flux. The 10 kD MWCO membrane removed approximately 4 logs of MS2 bacteriophage, while the two 100 kD MWCO membranes removed the bacteriophage to the sensitivity limit of the assay. The 100 kD membrane also removed PRD1 bacteriophage and poliovirus to their respective RSLs. The 300,000 kD MWCO membrane removed 4.3 logs of MS2 bacteriophage, and the 0.2 μm membrane, less than 0.5 log. A similar, but not as pronounced, removal trend was observed for PRD1 bacteriophage and PV1. The poorer removal by the mPS 10 kD membrane may have been due to a wide size distribution of pores or a few large pores; the manufacturer did not report pore size distribution data. No loss of membrane integrity was observed. The lower specific flux for this membrane suggests that it was "tighter" in construction than the 100 kD membranes. However, the data from these experiments support the hypothesis that manufacturer-reported MWCO or pore size and specific flux may not, in all cases, be a good predictor of virus removal.

Impact of Viral Loading

There are little data in the literature on whether exposing membranes to high numbers of specific organisms could cause "leakage" of the microorganisms into the filtrate. Experiments were conducted to evaluate the impact of viral loading on removal of microorganisms by three

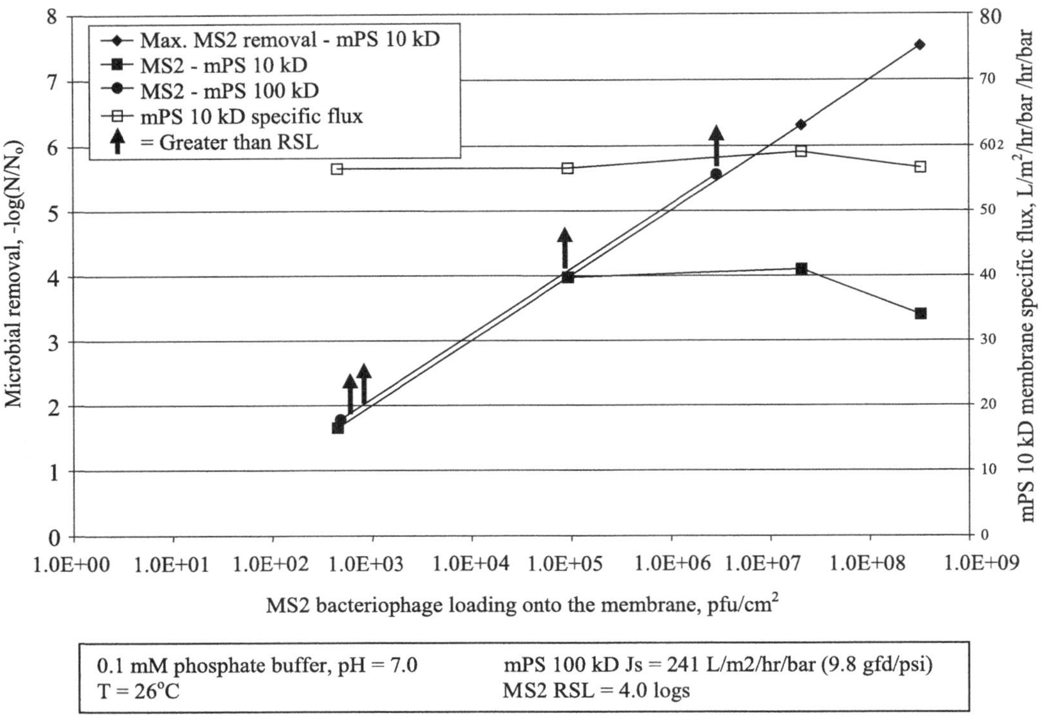

Figure 4.17 Removal of MS2 bacteriophage at various loadings onto two modified polysulfone membranes

selected membranes: mPS10 kD, mPS 100 kD and PVDF 0.035 μm (all as reported by the manufacturer). Each was exposed to various loadings of MS2 bacteriophage and removal was assessed. The data for the mPS membranes are shown in Figure 4.17. Specific flux is shown on the figure for the mPS10 kD membrane at each exposure point. Theoretical calculated removals based upon the bacteriophage being removed to the sensitivity limits of the assay are also presented. At loadings up to 10^5 pfu/cm^2 of membrane area, MS2 bacteriophage was removed to the sensitivity limits of the assay (4.0 logs) for both membranes. As the membrane exposure to bacteriophage increased beyond this point, mPS 100 kD membrane removed MS2 bacteriophage to the sensitivity limit and followed the theoretical removals. However, MS2 bacteriophage removal for the mPS 10 kD membrane remained constant or decreased. At a bacteriophage loading of 10^8 pfu/cm^2, only 3.4 logs of MS2 bacteriophage removal were observed. The data suggest that the loading of high levels of MS2 bacteriophage impacted the capability of the membrane to remove the bacteriophage.

A similar trend was observed with the PVDF 0.035 μm membrane, as shown in Figure 4.18. As viral loading increased from 1.6×10^6 pfu/cm^2 to 1.6×10^7 pfu/cm^2, MS2 bacteriophage removal decreased from 4.1 to 3.1 logs. However, the data in Figure 4.17 demonstrate that any observed "leakage" of virus from membranes is membrane specific. Further, it is important to note that the viral loadings in these experiments were extremely high and are unlikely to be

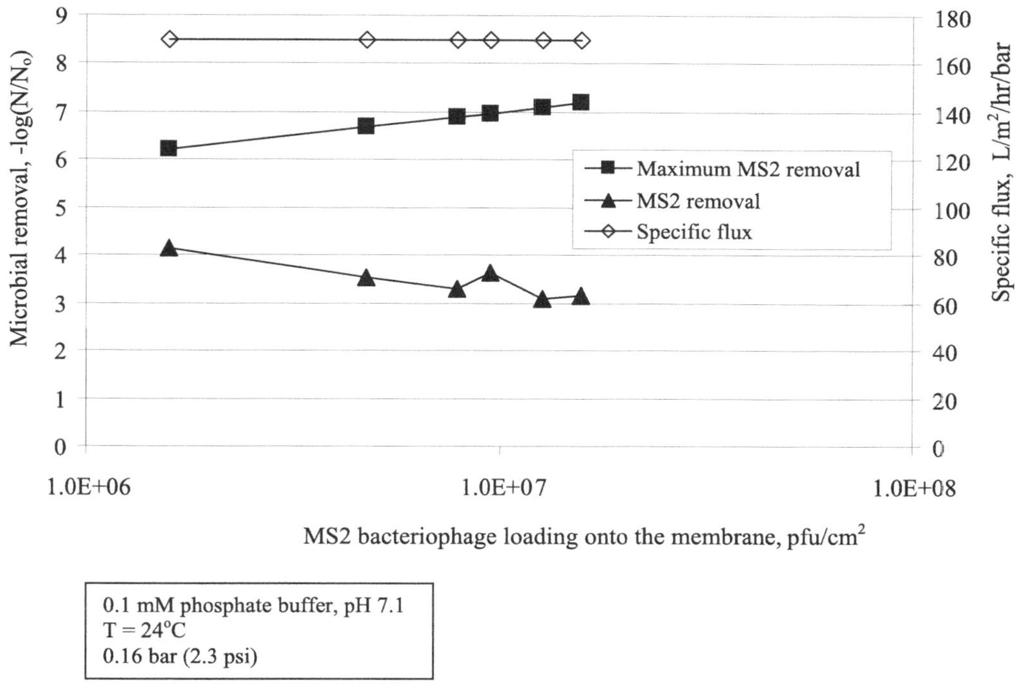

Figure 4.18 Removal of MS2 bacteriophage at various loadings onto PVDF 0.035 μm membrane

experienced in filtration of natural water supplies. For example, if one assumes the following conservative parameters for a full-scale membrane and its operation:

- active membrane area: 30 m^2
- membrane flux: 10 $L/m^2/hr$
- backwash frequency: 1 backwash every 2 hours
- concentration of viruses in source water: 10 virus/L

then the total viral loading on the membrane would only be 20 $virus/cm^2$ between backwashes. Thus, while viral loading may directly impact the viral filtrate quality from selected membranes, it is probable that this would not occur, even when treating severely impaired water sources.

BENCH-SCALE STUDIES - UNIVERSITY OF COLORADO, BOULDER

Bench-scale studies were conducted to evaluate the use of fluorescent microspheres as surrogates for bacteriophage in drinking water membrane studies and to evaluate the effects of compromising membranes on the rejection of microspheres and bacteriophage. The first step in these studies was to conduct a detailed evaluation of the characteristics of the bacteriophage, fluorescent microspheres and the membranes themselves in order to understand their interactions and the potential mechanisms of bacteriophage and microsphere removal. Secondly, experiments were

Table 4.4
Flux under clean water conditions for uncompromised flat-sheet membranes

Membrane	Clean water flux ($L/m^2/hr$) at 1 bar
RC-hpi	553
PES-hpo	345
PVDF-hpi	5,910
PVDF-hpo	476

performed to quantify the removal of bacteriophage and fluorescent microspheres by four different types of membranes. The four types of membranes were:

- Hydrophilic polyvinlyidine fluoride (designated "PVDF-hpi")
- Hydrophobic PVDF (designated "PVDF-hpo")
- Hydrophilic regenerated cellulose (designated "RC-hpi")
- Hydrophobic polyethersulfone (designated "PES-hpo")

The various conditions under which bacteriophage and microsphere removal were assessed included pH (6.5 and 8.5), presence or absence of fouling constituents (NOM and calcium) in the feedwater, and membrane compromising (uncompromised and compromised by one hole). Flat-sheet membranes with a 25-mm diameter were employed in a dead-end, stirred-cell bench-scale testing unit were used for all experiments.

Clean Water Flux of Uncompromised Membranes

Flux under clean water conditions is shown in Table 4.4. In all microsphere and bacteriophage challenges, the pressure was maintained at 1 bar (15 psi). However, as indicated in Table 4.4, the flux of each membrane was significantly different at that pressure. Thus, the particle loading velocity differed from membrane to membrane. The PVDF-hpi membrane was the highest flux membrane, followed by the RC-hpi, PVDF-hpo and PES-hpo membranes, in that order.

Effect of pH on Bacteriophage and Microsphere Removal

Microbial challenge tests with MS2 and PRD1 bacteriophage were performed at two different pH levels (6.5 and 8.5) using the four membranes, RC-hpi, PES-hpo, PVDF-hpi, and PVDF-hpo. The feedwater was deionized water having an ionic strength adjusted to 0.01 M using sodium chloride. Over this pH range, all four membranes were negatively charged, as were both bacteriophage. Thus, there should have been little change in the surface chemistry of the membranes or bacteriophage; presumably, this would result in little change in removal as a function of pH for each combination of bacteriophage and membrane. As shown in Figures 4.19 and 4.20, the difference in pH did not have a substantial effect on the removal of either PRD1 or MS2 bacteriophage by any of the membranes. These results are consistent with those of JHU bench testing, in which pH was not found to affect microbial removal within the pH range 5 to 9.

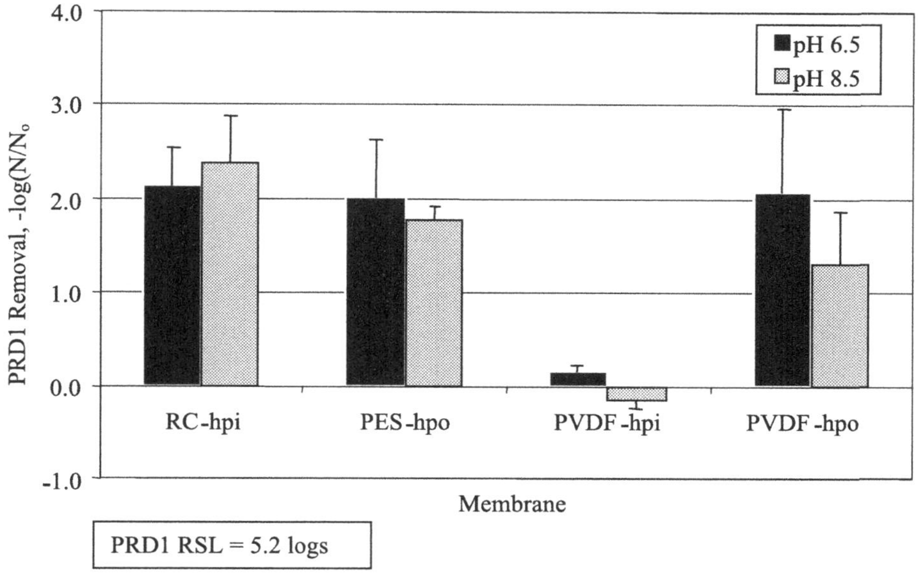

Figure 4.19 Effect of pH on PRD1 bacteriophage removal by flat-sheet membranes

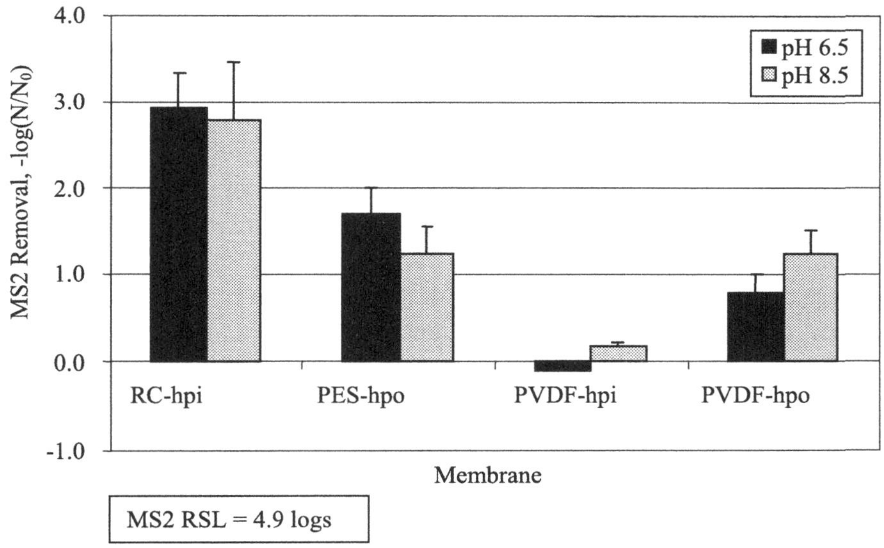

Figure 4.20 Effect of pH on MS2 bacteriophage removal by flat-sheet membranes

The above data at pH 6.5 were examined to contrast the bacteriophage removal performance of the two PVDF membranes versus the RC-hpi and PES-hpo membranes. PRD1 bacteriophage has a smaller diameter (70 nm) than the manufacturer-reported pore size of the two PVDF membranes tested (220 nm), but a larger diameter than the average MWCO of the RC or PES membranes tested (100 kD, calculated based on Equation 2.1 using MWCO reported by the manufacturer). Consequently, based on size exclusion, both of the PVDF membranes tested were expected to allow PRD1 bacteriophage to pass through, and the RC and PES membranes tested

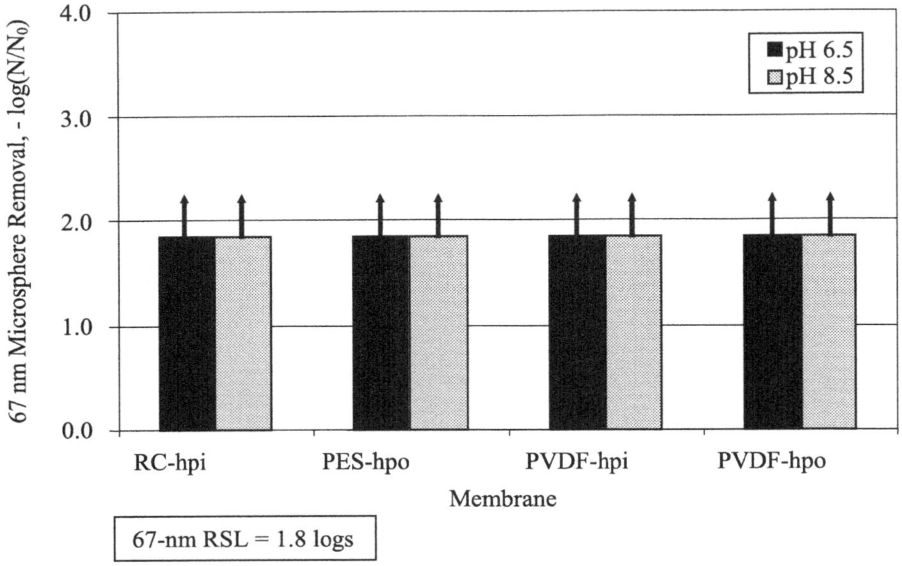

Figure 4.21 Effect of pH on removal of 67-nm spheres by flat-sheet membranes

were expected to retain PRD1 bacteriophage. The removal of PRD1 bacteriophage was consistent for the RC-hpi, PES-hpo, and PVDF-hpo membranes. The PVDF-hpi membrane, however, provided much lower PRD1 bacteriophage removal than the other three membranes. Similarly, MS2 bacteriophage removal results for the four membranes did not follow expected trends based on pore size. These results are consistent with those of JHU bench-scale testing, in which manufacturer-reported MWCO or pore size was not necessarily predictive of microbial removal. Further, the hydrophilic or hydrophobic nature of the membranes did not appear to impact microbial results in a consistent manner.

Microsphere challenge tests were performed at the same two pH levels and under the same conditions used in the microbial challenges described above. At these pH values, the four membranes were all negatively charged, as were both microspheres. Thus, little change was expected in the surface chemistry of the membranes or microspheres; presumably this would result in little change in removal for each combination of microsphere and membrane at the two different pH levels. As shown in Figure 4.21, the difference in pH did not have a significant effect on the removal of 67-nm spheres by each of the membranes. This was due in part to the low sensitivity limit of the assay for 67-nm microspheres (1.8 logs), and the resulting inability to distinguish trends in removal. As shown in Figure 4.22, removal of the 26-nm spheres (with a sensitivity limit of 3.1 logs) was lower by 1.0 to 1.5 logs at pH 8.5 than at pH 6.5.

As observed above with respect to microbial removal, microsphere removal was not found to be consistent among the membranes based on size exclusion or the hydrophilic or hydrophobic nature of the membranes. The similar removal of the 67-nm microspheres by all four membranes (Figure 4.21) is explained in part by the very low sensitivity limit of the assay (less than 2.0 logs).

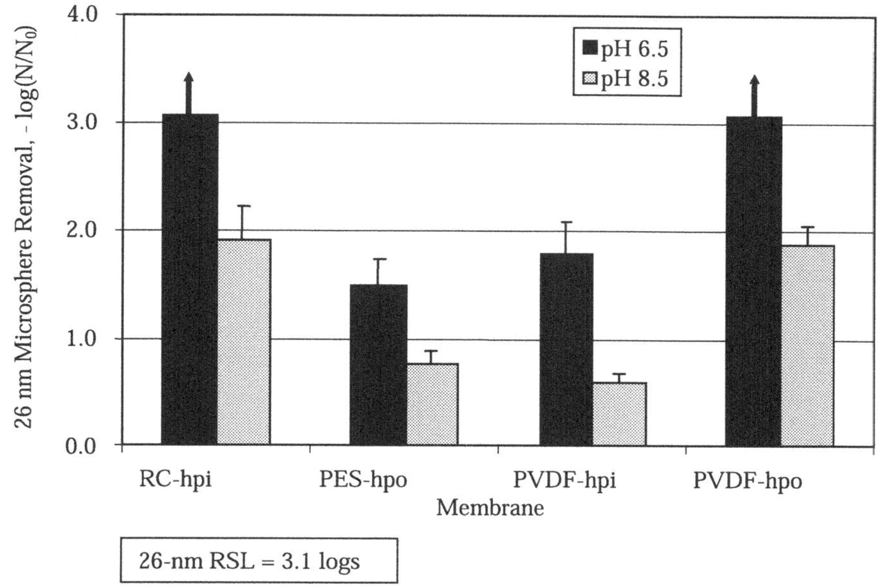

Figure 4.22 Effect of pH on removal of 26-nm spheres by flat-sheet membranes

Comparison of Results for MS2 Bacteriophage and 26-nm Microspheres

The results for MS2 and 26-nm microspheres were compared at pH 6.5 to evaluate potential differences in removal mechanisms. As shown in Figure 4.23, little difference was observed between removal of the bacteriophage and microspheres with the RC or PES membranes. However, in the case of the two PVDF membranes, removal of the 26-nm microspheres was much higher than MS2 removal. Because the pore size of the RC and PES membranes tested was smaller than the diameter of both the 26-nm microspheres and the MS2, removal by the RC and PES membranes should have been based only on size exclusion, which explains why little difference was observed in results for the 26-nm spheres and MS2. However, the difference between bacteriophage and microsphere removal by the two PVDF membranes, which had nominal pore sizes larger than both the 26-nm spheres and MS2 is not as readily explained.

The effect of the presence of fouling constituents in the feedwater on removal of 26-nm spheres and MS2 was compared. Fouling experiments were conducted on both of the PVDF membranes at pH 6.5. Fouling constituents added to the feedwater were comprised of 5 mg/L of Suwannee River fulvic acid NOM (0.03 mg C/cm^2 delivered) and 100 mg/L of calcium added as calcium chloride. The presence of the fouling constituents caused increased removal of 26-nm spheres by the PVDF-hpi membrane, but did not lead to improved removal of the spheres by the PVDF-hpo membrane, or improved removal of MS2 by either of the PVDF membranes tested. Figure 4.24 also indicates that regardless of the presence of fouling constituents, microsphere removal by both of the PVDF membranes tested was consistently higher than that of MS2.

Comparison of Results for PRD1 Bacteriophage and 67-nm Microspheres

Because pH effects were not observed for either bacteriophage or 67-nm microspheres, the results for PRD1 and 67-nm microspheres were compared at pH 6.5 only. As shown in

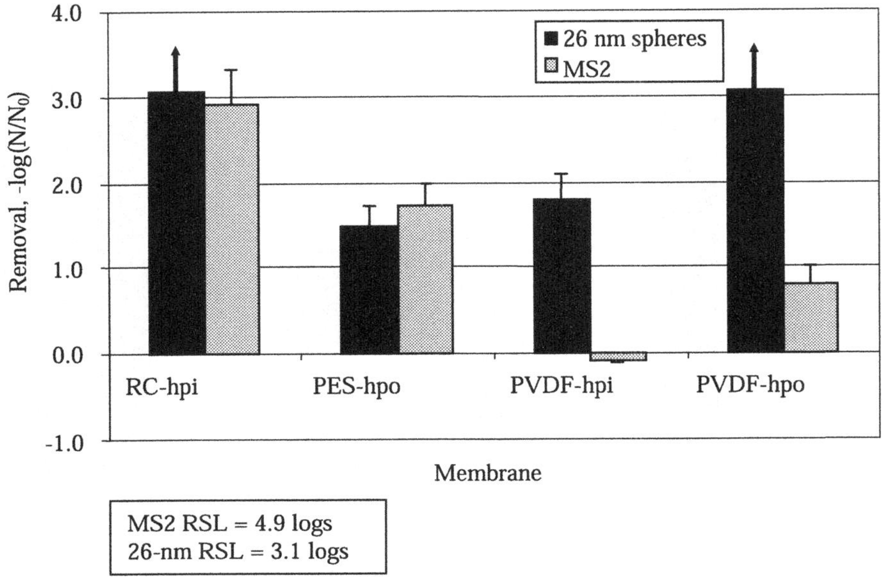

Figure 4.23 Removal of 26-nm spheres and MS2 bacteriophage at pH 6.5 by flat-sheet membranes

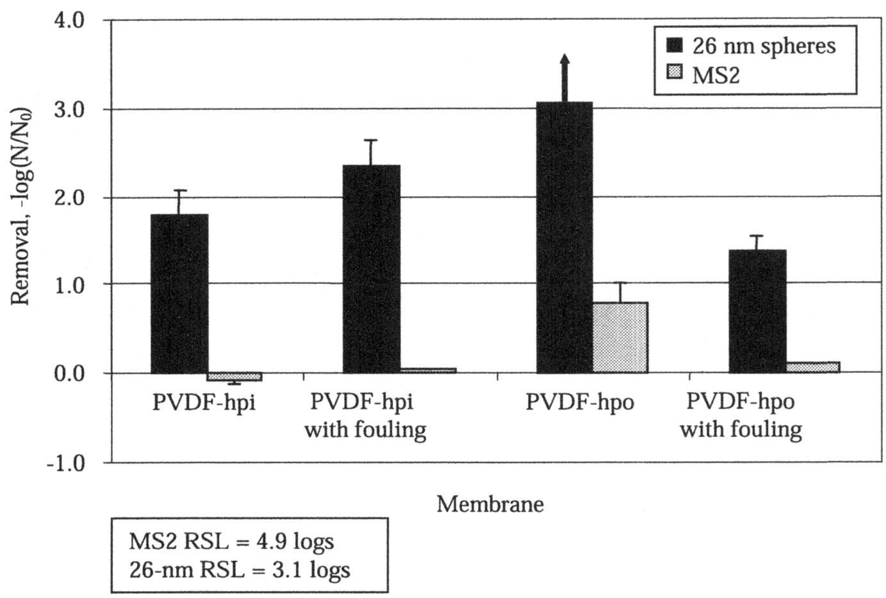

Figure 4.24 Effect of the presence of fouling constituents in the feedwater on removal of 26-nm spheres and MS2 bacteriophage by flat-sheet membranes

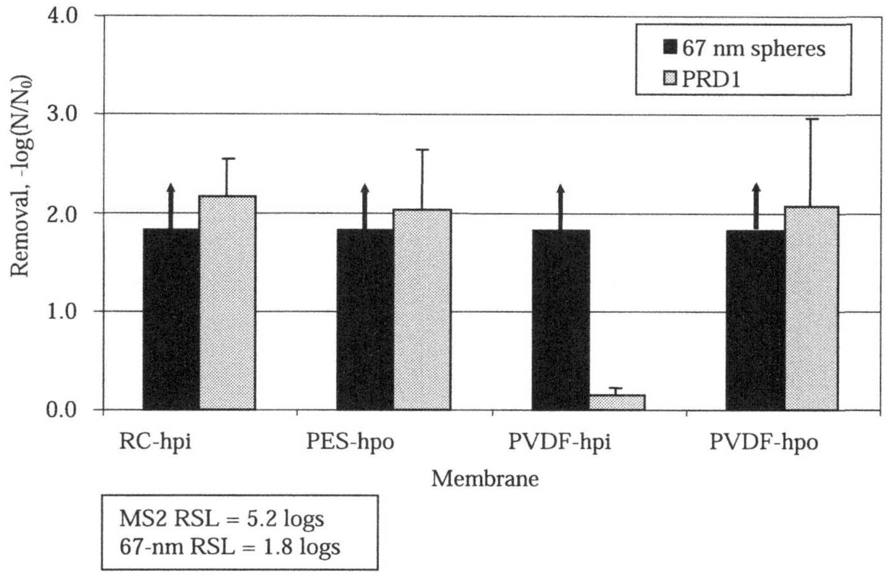

Figure 4.25 Removal of 67-nm spheres and PRD1 bacteriophage at pH 6.5 by flat-sheet membranes

Figure 4.25, little difference was observed between removal of PRD1 and 67-nm microspheres by the RC, PES and PVDF-hpo membranes tested. However, the low sensitivity limit of the 67-nm sphere assay should be noted; thus, clear differences could not be discerned. In the case of the PVDF-hpi membrane, the 67-nm microspheres were removed much more effectively than PRD1. This difference in removal cannot be explained by size exclusion, because the nominal pore size determined for both of the PVDF membranes (220 nm) was larger than that of both the spheres (67 nm) and PRD1 (70 nm).

The effects of the presence of fouling constituents in the feedwater on removal of 67-nm spheres and PRD1 by the two PVDF membranes were also compared under the same conditions described above for 26-nm spheres and MS2. Figure 4.26 indicates that the presence of fouling constituents decreased removal of PRD1 by the PVDF-hpo membrane, a trend that was also observed in removal of MS2 by this membrane (Figure 4.24). Further, Figure 4.24 shows that microsphere removal by three of the four membranes tested was much more effective than that of PRD1. For the PVDF-hpo membrane, microsphere and PRD1 removal were similar. However, this observation was due in part to the low sensitivity limit of the 67-nm sphere assay (1.8 logs).

Membrane Integrity

As described in Chapter 2 Methods and Materials, membranes were compromised using a needle, which punctured a hole on the order of 100 μm in diameter. The surface area of a 100-μm hole was on the order of 0.005% of the area of the flat-sheet membranes used in the bench-scale testing. Because hollow-fiber membrane systems are the predominant form of commercially available low-pressure membrane systems, it is important to compare the compromising of a

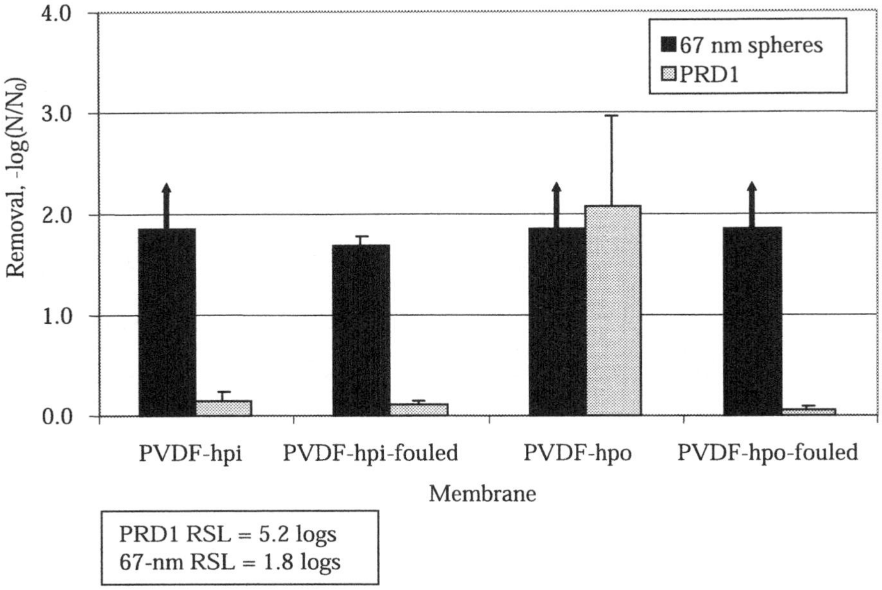

Figure 4.26 Effect of the presence of fouling constituents in the feedwater on removal of 67-nm spheres and PRD1 bacteriophage by flat-sheet membranes

flat-sheet membrane in this study with a pin to the breaking of a single fiber in a hollow-fiber module. Based on surface area alone, it was calculated that a hole of 0.005% of the membrane surface equates to one in ten fibers being broken. Thus, results should be interpreted with this comparison in mind, as this is a large number of fibers to be compromised.

In all cases, compromising the membrane with one hole of a diameter on the order of 100 μm substantially decreased bacteriophage removal. An example is provided in Figure 4.27 for PRD1 removal by the four membranes tested, at two different pH levels (6.5 and 8.5). Comparing the results in Figure 4.27 for the compromised membranes to those in Figure 4.22 for uncompromised membranes, it is clear that one hole of approximately 100 μm comprising only 0.005% of the membrane surface area was sufficient to drastically decrease PRD1 removal. The same trend was observed comparing uncompromised membranes to membranes compromised by one hole for all four membranes and removal of MS2, 67-nm microspheres and 26-nm microspheres under all conditions tested (pH of 6.5 and 8.5, and presence or absence of fouling materials; data not shown).

The time required for the sample to pass through the membrane was much lower for a compromised membrane than for the corresponding uncompromised membrane. The substantial differences in flux observed in the bench-scale testing between uncompromised and compromised membranes is reported in Table 4.5. Lytle and Routson (1995) state that it is fluid flow, and not virus size, that limits virus penetration through a single hole. Thus, under the conditions employed in this study, it is probable that much of the water flow passed through the hole instead of through the pores of the membrane, and carried the bacteriophage with it.

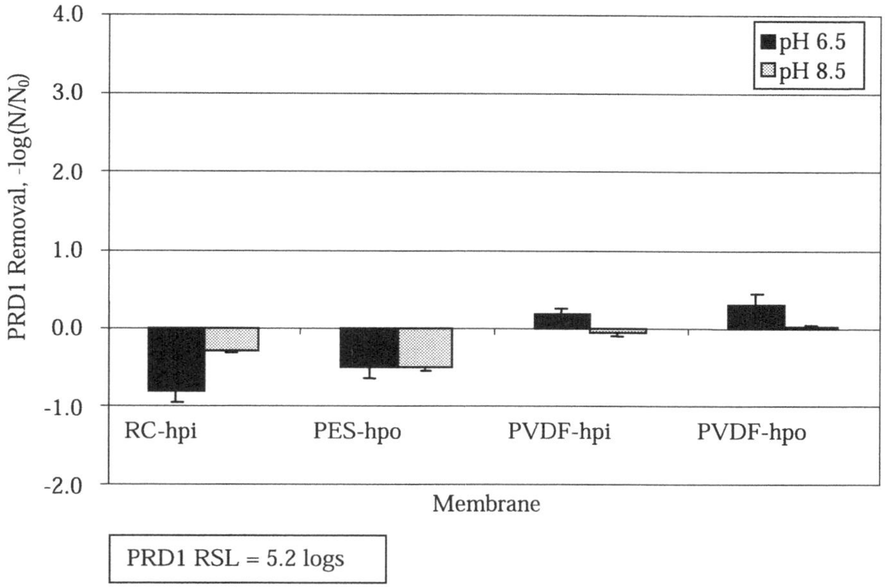

Figure 4.27 Effect of membrane compromising (1 hole) on PRD1 bacteriophage removal at two pH levels by flat-sheet membranes

Table 4.5
Flux of uncompromised and compromised flat-sheet membranes

	Clean water flux (L/m²/hr) at 1 bar	
Membrane	Uncompromised membrane (0 holes)	Compromised membrane (1 hole)
RC-hpi	553	5,590
PES-hpo	345	2,280
PVDF-hpi	5,910	9,750
PVDF-hpo	476	4,170

PILOT-SCALE STUDIES - UNITED STATES

Pilot-Scale Study at the Atlanta Fulton County Water Treatment Plant, Georgia

Pilot-scale testing at the Atlanta Fulton County Water Treatment Plant (AFC WTP) consisted of executing Tasks 1 through 4 of the developed protocol (see Chapter 2, Methods and Materials) as follows:

- Task 1: Hydraulic characterization
- Task 2: Evaluation of microbial removal under clean water conditions
- Task 3: Evaluation of the impact of deposition layer formation on microbial removal
- Task 4: Evaluation of the impact of membrane fouling on microbial removal

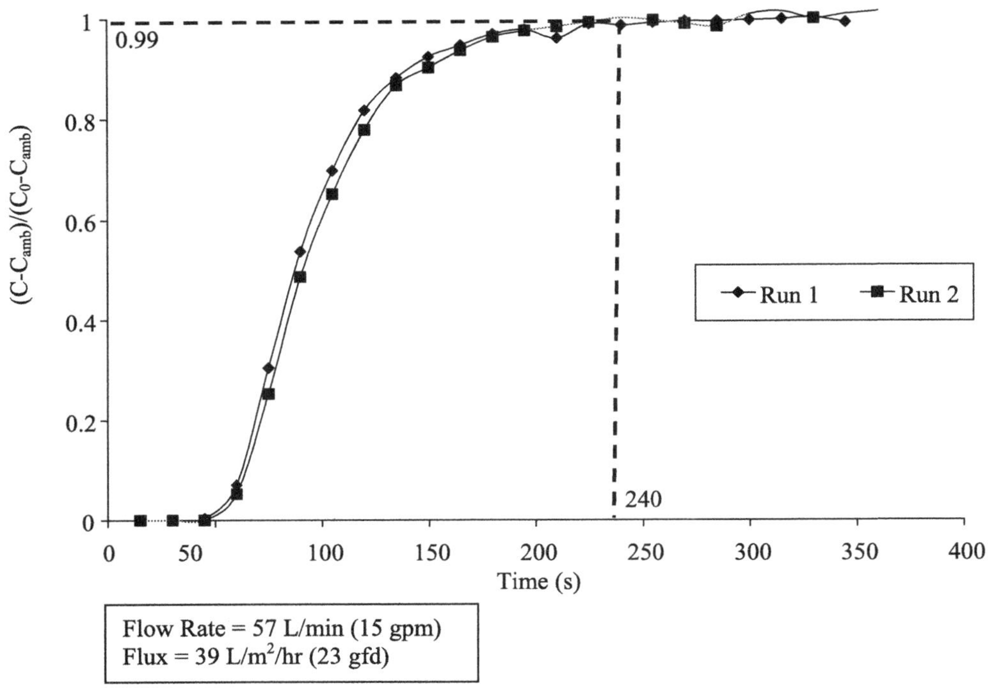

Figure 4.28 Tracer testing results at 39 L/m²/hr (pressurized PVDF membrane pilot unit, Atlanta)

These tasks were performed at the AFC WTP with both raw and filtered water using a PVDF pressurized membrane pilot unit, and with raw water using a PVDF submerged membrane pilot unit. Water quality for these two supplies was summarized in Chapter 2, Methods and Materials.

It should be noted that the submerged PVDF membrane used in this study was brand new at the time testing in Atlanta was initiated. The pressurized PVDF membrane used in this study had been in operation for some time at other locations with natural waters used as the feedwater. Thus, the maturity, or condition of irreversible fouling, of the PVDF membrane used in the pressurized pilot unit was greater than that of the brand-new PVDF membrane used in the submerged pilot unit.

Experimentation With the Pressurized PVDF Membrane Pilot Unit

Hydraulic Characterization. Tracer tests were conducted at four different flow rates with the pressurized PVDF pilot unit to determine the hydraulic residence time distribution. Each test was performed in duplicate. Figure 4.28 shows the results of the two tracer tests conducted at 57 L/min (15 gpm). Figure 4.29 depicts a linear regression of the T_{99} value as a function of the flow rate, demonstrating that the T_{99} for the pressurized PVDF pilot unit was inversely proportional to the flow rate (for flow rates ranging from 38 to 76 L/min). Figure 4.30 shows the number of HRTs required to reach T_{99}. For all the tested flow conditions, steady-state conditions were reached after approximately 2.5 HRTs. The findings from bench-scale studies indicated that 5 HRTs were necessary to achieve steady-state conditions. To be consistent, the first filtrate samples were collected at 5 HRTs after initiating the microbial seeding.

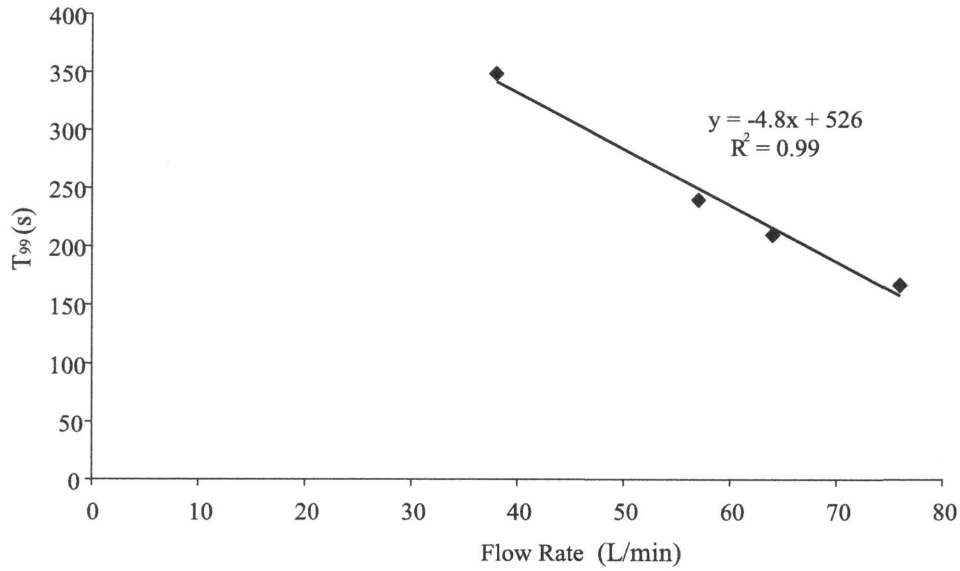

Figure 4.29 T_{99} as function of flow rate (pressurized PVDF membrane pilot unit, Atlanta)

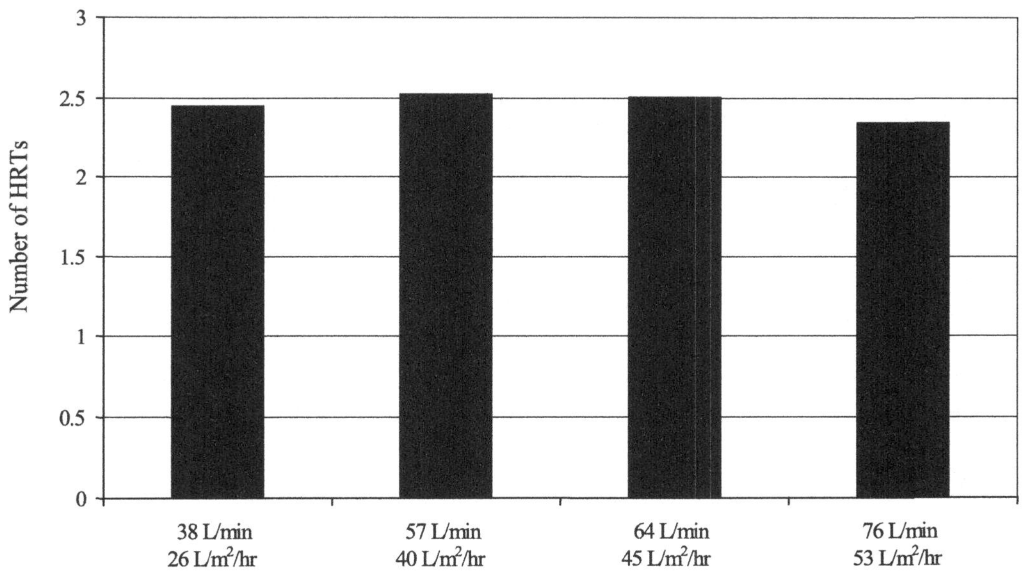

Figure 4.30 Number of HRTs to reach T_{99} (pressurized PVDF membrane pilot unit, Atlanta)

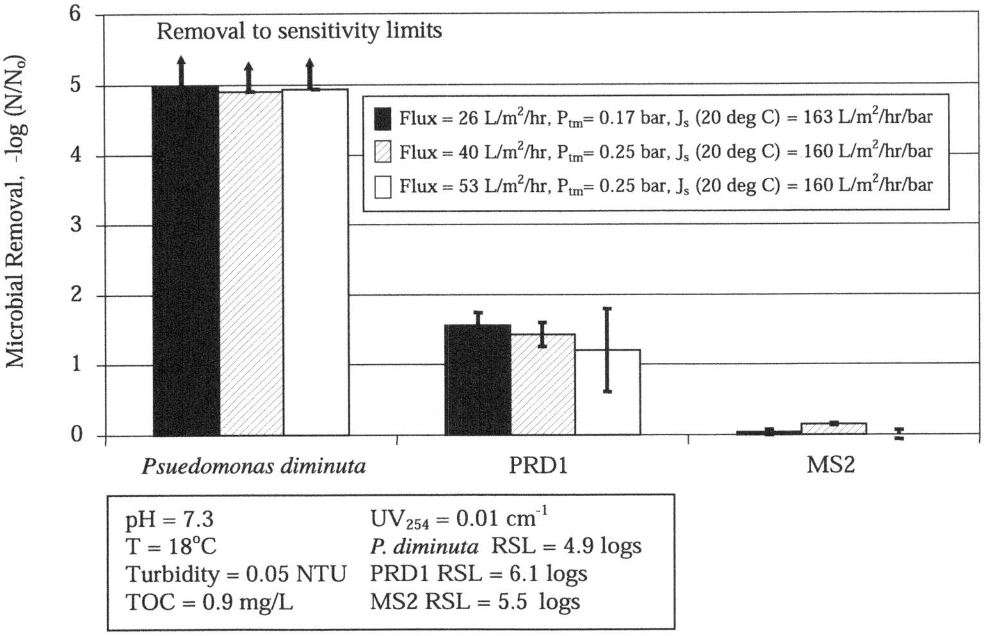

Figure 4.31 Microbial removal under clean water conditions (pressurized PVDF membrane pilot unit, Atlanta)

Microbial Removal Efficiency Under Clean Water Conditions. Two microbial challenges were conducted with the PVDF pressurized pilot unit using MS2 and PRD1 bacteriophage and Pseudomonas diminuta seeded into dechlorinated tap water (finished water from the AFC WTP). Microbial challenges were performed at three different fluxes within the same filtration cycle (i.e., with no backwashing).

Figure 4.31 shows the results obtained for *P. diminuta*, MS2 bacteriophage (0.026 μm) and PRD1 bacteriophage (0.065 μm) removal by the pressurized PVDF pilot unit. Error bars were calculated based on the experimental standard deviation of results from duplicate seeding challenges at each condition. For all three operating conditions, the removal of PRD1 bacteriophage was slightly higher than 1 log, while very little removal of MS2 bacteriophage was observed. Filtration operating conditions (e.g., flux) did not influence bacteriophage removal.

As expected, based on the manufacturer-reported nominal pore size (0.1 μm) of the membrane and the size of *Pseudomonas diminuta* (0.6 to 1.0 μm), the pressurized PVDF pilot unit achieved removal to the sensitivity limit of the microbial assay.

Confidence intervals (95%) were calculated around the mean for each of the results shown in Figure 4.31 (error bars, not confidence intervals are shown on graph). For *P. diminuta*, PRD1 and MS2 the confidence intervals ranged from 4.3 to 5.6 logs, 0.04 to 2.4 logs, and -0.03 to 0.2 logs, respectively. Thus, while there was a statistically significant difference between the removal of *P. diminuta* and the two bacteriophage, the difference between removal of PRD1 and MS2 under clean water conditions was not statistically significant.

Operating with clean water is the most favorable condition for microbial passage through the membrane system as compared to a fouled system, where mechanisms such as deposition layer formation and organic and/or inorganic fouling can enhance microbial removal. As removal

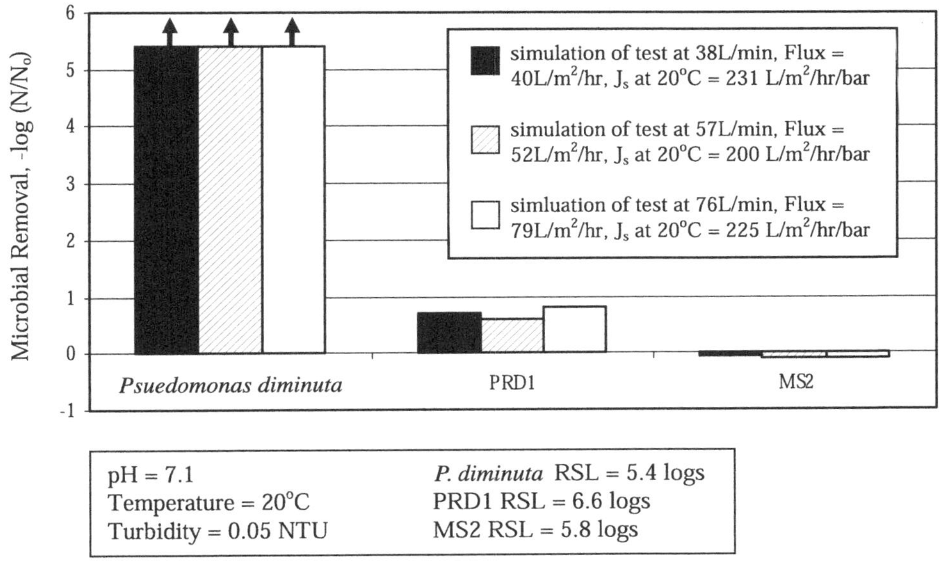

Figure 4.32 Microbial removal observed in the PVDF membrane bench-scale unit under clean water conditions

to the sensitivity limits of the assay for *P. diminuta* was observed under clean water conditions, it was decided to focus only on MS2 and PRD1 bacteriophage for subsequent testing with the pressurized PVDF pilot unit using raw and filtered water from the AFC WTP. MS2 and PRD1 bacteriophage removal under clean water conditions was used for calculating the net microbial removal under conditions of deposition layer formation and membrane fouling (see Equation 2.5 in Chapter 2, Methods and Materials).

A comparison between the microbial removal observed at bench and pilot scale was conducted. Samples of dechlorinated tap water used for pilot-scale experiments were shipped to JHU and used to conduct microbial challenges with MS2 and PRD1 bacteriophage and *P. diminuta* with a bench-scale unit equipped with similar membranes (the same type of PVDF hollow fiber membrane from the same manufacturer, designated "PVDF 0.1 μm" in bench testing). Figure 4.32 depicts the microbial removal obtained during bench-scale tests.

The results obtained at bench scale were comparable to the pilot-scale results. No removal of MS2 bacteriophage was observed, while *P. diminuta* was completely retained by the PVDF membrane system. Results from Figures 4.31 and 4.32 indicate that the PVDF membrane pilot unit provided slightly higher removal of PRD1 bacteriophage compared to the PVDF membrane bench-scale unit (removal of 1.2 to 1.6 logs with the pilot unit compared to removal of 0.5 to 0.7 logs with the bench-scale unit); however, these differences were small.

Evaluation of the Impact of Deposition Layer Formation on Microbial Removal. In Atlanta pilot testing of the pressurized PVDF membrane pilot unit, the impact of deposition layer formation was evaluated for both raw and filtered water from the AFC WTP.

Raw Water. Microbial challenge experiments were conducted on the pressurized PVDF membrane pilot unit at fluxes of 26, 40, and 53 L/m^2/hr (15, 24, and 31 gfd) with MS2 and PRD1 bacteriophage seeded into the raw water during a 4-hour run time with no backwashing. Raw water quality during these experiments is shown in Table 4.6. Figure 4.33 depicts the net microbial removal

Table 4.6
Raw water quality during deposition layer formation challenges (pressurized PVDF membrane pilot unit, Atlanta)

Parameter	Unit	Flow rate 38 L/min Flux 26 L/m^2/hr	Flow rate 57 L/min Flux 40 L/m^2/hr	Flow rate 76 L/min Flux 53 L/m^2/hr
Temperature	°C	18	18	18
Turbidity	NTU	3.0	2.5	2.2
pH		6.9	7.0	6.9
Hardness	mg/L $CaCO_3$	16	16	15
Alkalinity	mg/L $CaCO_3$	15	17	16
TOC	mg/L	1.3	1.0	1.5
TSS	mg/L	<2	<2	<1
UV_{254}	cm^{-1}	0.035	0.029	0.028

of MS2 and PRD1 bacteriophage at six sampling times during the filtration cycle for a constant flux of 26 L/m^2/hr (15 gfd). The decline in specific flux during this experiment is also shown.

With a decrease in specific flux of 13% after 240 minutes of filtration (9 m^3 of water filtered), little influence on bacteriophage removal was observed. The net removal of PRD1 bacteriophage ranged between 0.6 and 1.0 log, whereas the net removal of MS2 bacteriophage varied between 0.1 and 0.7 log. Similar results were observed at higher filtration fluxes (40 and 53 L/m^2/hr; 24 and 31 gfd). Table 4.7 summarizes the additional removal (compared to clean water conditions) observed for the three fluxes tested. The low TOC and particulate matter concentrations in the raw water (TOC = 1.0 to 1.5 mg/L, turbidity = 2.2 to 3.0 NTU) did not allow for the accumulation of a deposition layer with enough thickness and/or compaction to observe an increase in removal of MS2 and PRD1 bacteriophage over an extended filtration cycle in the pressurized PVDF membrane pilot unit.

Filtered Water. Microbial challenge experiments were conducted at fluxes of 26, 40 and 53 L/m^2/hr (15, 24 and 31 gfd) with MS2 and PRD1 bacteriophage seeded into dechlorinated filtered water from the AFC WTP during a single filtration cycle in the pressurized PVDF membrane pilot unit. Water quality parameters during these experiments are shown in Table 4.8.

Figure 4.34 depicts the net removal of MS2 and PRD1 bacteriophage at six sampling times during a single filtration cycle for a constant flux of 40 L/m^2/hr (24 gfd). The decline in specific flux during this experiment is also shown. MS2 bacteriophage removal was as low as that observed during the clean water microbial challenge (0.1 log). Additional PRD1 bacteriophage removal ranged between 0.0 and 1.3 logs, with the highest removal observed at the end of the filtration cycle. Due to the low particulate and NOM content of the filtered water (turbidity between 0.02 and 0.10 NTU, and TOC between 1.0 and 1.4 mg/L), the specific flux decline during the 4-hour filtration cycle was very low, demonstrating the absence of a significant deposition layer on the membrane. Therefore, the microbial removal observed during a filtration cycle with filtered water was similar to that observed under clean water conditions. The results at the other

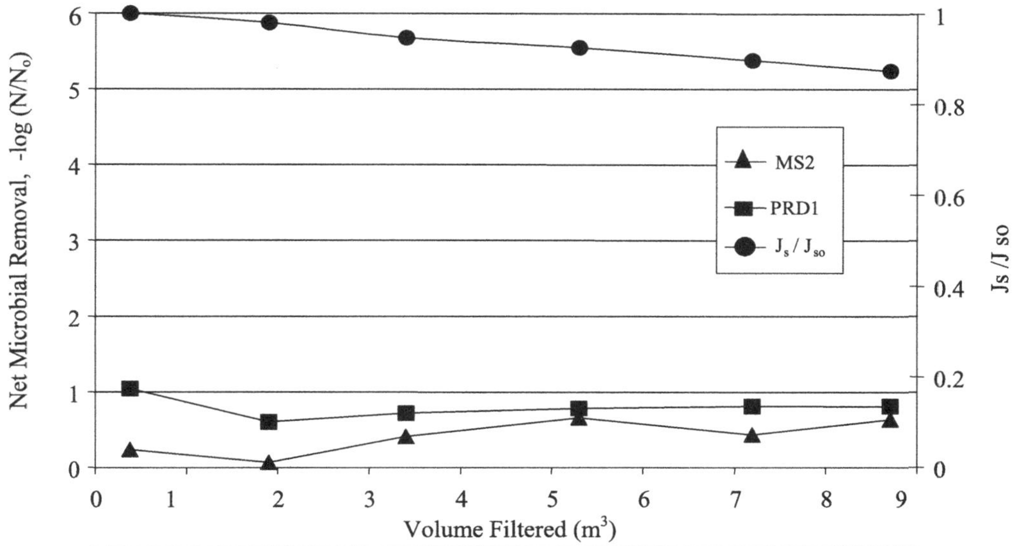

Flux = 26 L/m²/hr (16 gfd) | J_s (20°C) = 167 L/m²/hr/bar (6.8 gfd/psi)
Flow Rate = 38 L/min (10 gpm) | MS2 RSL = 5.8 logs
P_{tmo} = 0.16 bar (2.3 psi) | PRD1 RSL = 6.3 logs

Note:
MS2 bacteriophage removal under clean water conditions at 26 L/m^2/hr = 0.0 log
PRD1 bacteriophage removal under clean water conditions at 26 L/m^2/hr = 1.6 logs

Figure 4.33 Impact of deposition layer formation on microbial removal (raw water, pressurized PVDF membrane pilot unit, flux = 26 L/m^2/hr, Atlanta)

Table 4.7
Additional microbial removal (compared to clean water conditions) observed during a 4-hour filtration cycle (raw water, pressurized PVDF membrane pilot unit, Atlanta)

	Flow rate = 38 L/min Flux = 26 L/m^2/hr	Flow rate = 57 L/min Flux = 40 L/m^2/hr	Flow rate = 76 L/min Flux = 53 L/m^2/hr
Total volume filtered	9 m^3	14 m^3	18 m^3
Reduction in specific flux after 4 hours of filtration	13%	19%	14%
Additional removal of MS2 bacteriophage	0.1 to 0.7 log	0 to 1.3 log	0.2 to 0.8 log
Additional removal of PRD1 bacteriophage	0.6 to 1.0 log	0.8 to 1.3 log	0.6 to 1.3 log

Table 4.8
Filtered water quality during deposition layer formation challenges (pressurized PVDF membrane pilot unit, Atlanta)

Parameter	Unit	Flow rate 38 L/min Flux 26 $L/m^2/hr$	Flow rate 57 L/min Flux 40 $L/m^2/hr$	Flow rate 76 L/min Flux 53 $L/m^2/hr$
Temperature	°C	13	12	11
Turbidity	NTU	0.02	0.03	0.10
pH		6.2	6.2	6.2
Hardness	mg/L $CaCO_3$	19	18	19
Alkalinity	mg/L $CaCO_3$	12	12	18
TOC	mg/L	1.4	1.3	1.0
TSS	mg/L	<1	<1	<2
UV_{254}	cm^{-1}	0.024	0.022	0.020

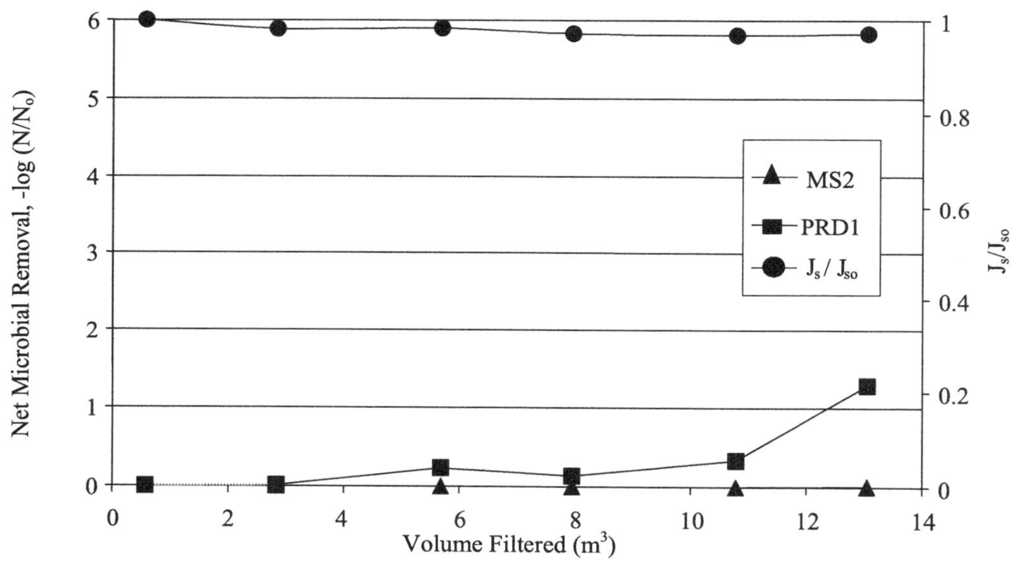

Flux = 40 $L/m^2/hr$ (23 gfd)	J_{so} (20°C) = 184 $L/m^2/hr/bar$ (7.5 gfd/psi)
Flow Rate = 57 L/min (15 gpm)	MS2 RSL = 7.6 logs
P_{tmo} = 0.26 bar (3.8 psi)	PRD1 RSL = 6.7 logs

Note:
MS2 bacteriophage removal under clean water conditions at 40 $L/m^2/hr$ = 0.2 log
PRD1 bacteriophage removal under clean water conditions at 40 $L/m^2/hr$ = 1.4 logs

Figure 4.34 Impact of deposition layer formation on microbial removal (filtered water, pressurized PVDF membrane pilot unit, flux = 40 $L/m^2/hr$, Atlanta)

Table 4.9
Additional microbial removal (compared to clean water conditions) observed during a 4-hour filtration cycle (filtered water, pressurized PVDF membrane pilot unit, Atlanta)

	Flow rate = 38 L/min Flux = 26 $L/m^2/hr$	Flow rate = 57 L/min Flux = 40 $L/m^2/hr$	Flow rate = 76 L/min Flux = 53 $L/m^2/hr$
Total volume filtered	9 m^3	14 m^3	18 m^3
Reduction in specific flux after 4 hours of filtration	<1%	3%	3%
Additional removal of MS2 bacteriophage	<0.4 log	<0.1 log	<0.1 log
Additional removal of PRD1 bacteriophage	<0.5 log	0 to 1.3 log	0 to 0.7 log

fluxes, 26 $L/m^2/hr$ and 53 $L/m^2/hr$ (15 and 31 gfd), are summarized in Table 4.9. These results confirm the negligible short-term impact of the particulate and organic content of the filtered water on bacteriophage removal under the test conditions.

Evaluation of the Impact of Membrane Fouling on Microbial Removal. In Atlanta pilot testing of the pressurized PVDF membrane pilot unit, the impact of membrane fouling was evaluated for both raw and filtered water from the AFC WTP.

Raw Water. For purposes of these experiments, membrane fouling was defined as the decline in specific flux that was not attributable to a backwashable deposition layer. The pressurized PVDF membrane pilot unit was operated for a two-week period at a constant flux of 40 $L/m^2/hr$ (24 gfd) with a backwash interval of 15 minutes. Water quality during this experiment is reported in Table 4.10. The water quality remained stable over the course of the experiment. Six microbial challenges were conducted during the two weeks of testing.

Figure 4.35 depicts the net removal of MS2 and PRD1 bacteriophage as a function of filtration time. Over the two weeks of operation (11.5 days of effective filtration), the specific flux declined by 58%. The net removal of PRD1 bacteriophage ranged between 2.7 and 4.9 logs. While the removal of PRD1 bacteriophage gradually increased over the testing period, the results obtained with MS2 bacteriophage were not as consistent as those of PRD1 (additional removal varying between 0.2 and 5.0 logs).

Filtered Water. Testing was conducted in the pressurized PVDF membrane pilot unit over a two-week period at a flux of 53 $L/m^2/hr$ (31 gfd) with a backwash interval of 90 minutes. Due to the lower particulate and NOM content of the filtered water compared to the raw water, a higher flux and a lower backwash frequency were selected to produce a significant specific flux decrease over the two-week experiment with filtered water. Water quality during this experiment is reported in Table 4.11. The water quality remained stable over the course of the experiment. Six microbial challenges were conducted over the two weeks of testing.

Figure 4.36 depicts the net removal of MS2 and PRD1 bacteriophage as a function of filtration time during the two-week test. Over the two weeks of operation (13.6 days of effective filtration), the specific flux declined by 31%. MS2 bacteriophage removal was not significantly

Table 4.10
Raw water quality during membrane fouling challenges
(pressurized PVDF membrane pilot unit, Atlanta)

Parameter	Unit	Average (min - max)
Turbidity	NTU	2.9 (2.4-3.4)
Hardness	mg/L $CaCO_3$	16 (15-17)
Alkalinity	mg/L $CaCO_3$	16 (15-17)
Temperature	°C	17 (16-17)
TOC	mg/L	1.5 (1.2-1.7)
TSS	mg/L	1.7 (<1-1.9)
pH		6.8 (6.7-6.8)

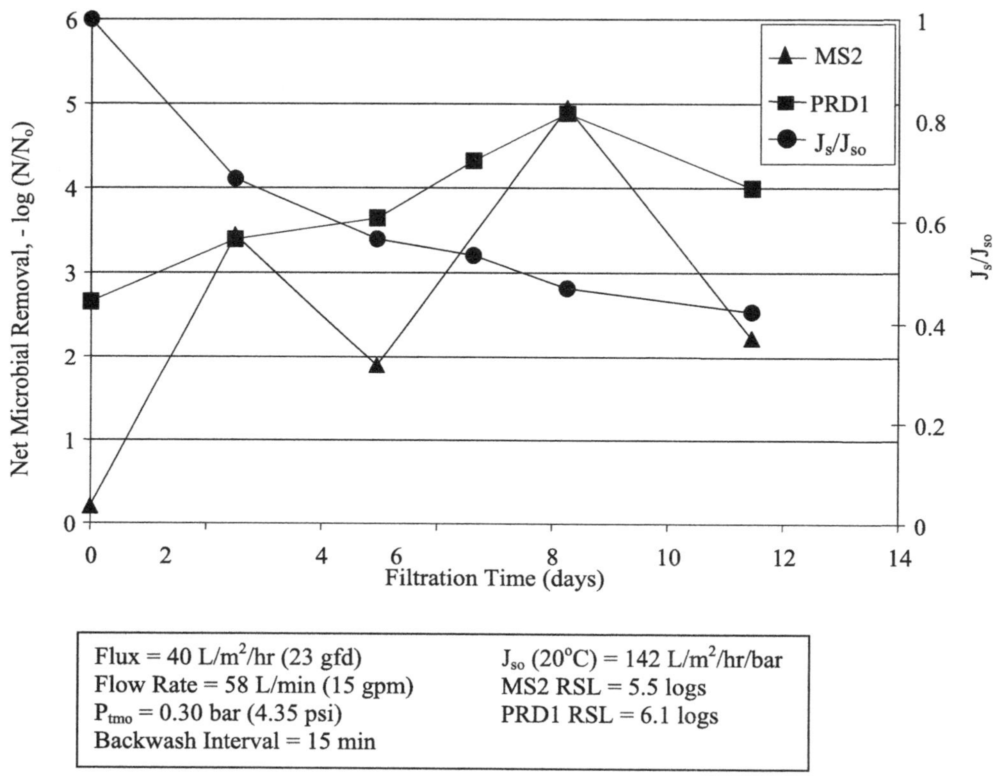

Note:
MS2 bacteriophage removal under clean water conditions at 40 L/m^2/hr = 0.2 log
PRD1 bacteriophage removal under clean water conditions at 40 L/m^2/hr = 1.4 logs

Figure 4.35 Impact of fouling on microbial removal over a two-week filtration experiment (raw water, pressurized PVDF membrane pilot unit, flux = 40 L/m^2/hr, Atlanta)

Table 4.11
Filtered water quality during membrane fouling challenges (pressurized PVDF membrane pilot unit, Atlanta)

Parameter	Unit	Average (min-max)
Turbidity	NTU	0.05 (0.02-0.06)
Hardness	mg/L $CaCO_3$	16 (15-16)
Alkalinity	mg/L $CaCO_3$	14.4 (12-16)
Temperature	°C	7 (7-8)
TOC	mg/L	1.3 (1.1-1.4)
TSS	mg/L	<1
pH		6.5 (6.4-6.5)

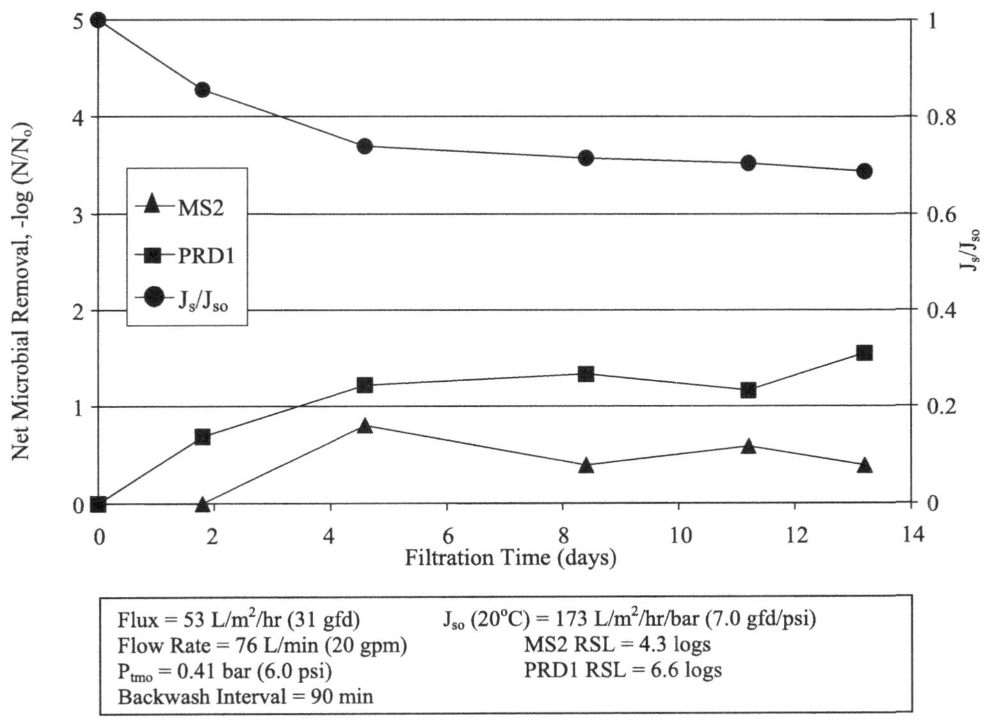

Note:
MS2 bacteriophage removal under clean water conditions at 53 L/m^2/hr = 0.0 log
PRD1 bacteriophage removal under clean water conditions at 53 L/m^2/hr = 1.2 logs

Figure 4.36 Impact of fouling on microbial removal over a two week filtration experiment (filtered water, pressurized PVDF membrane pilot unit, flux = 53 L/m^2/hr, Atlanta)

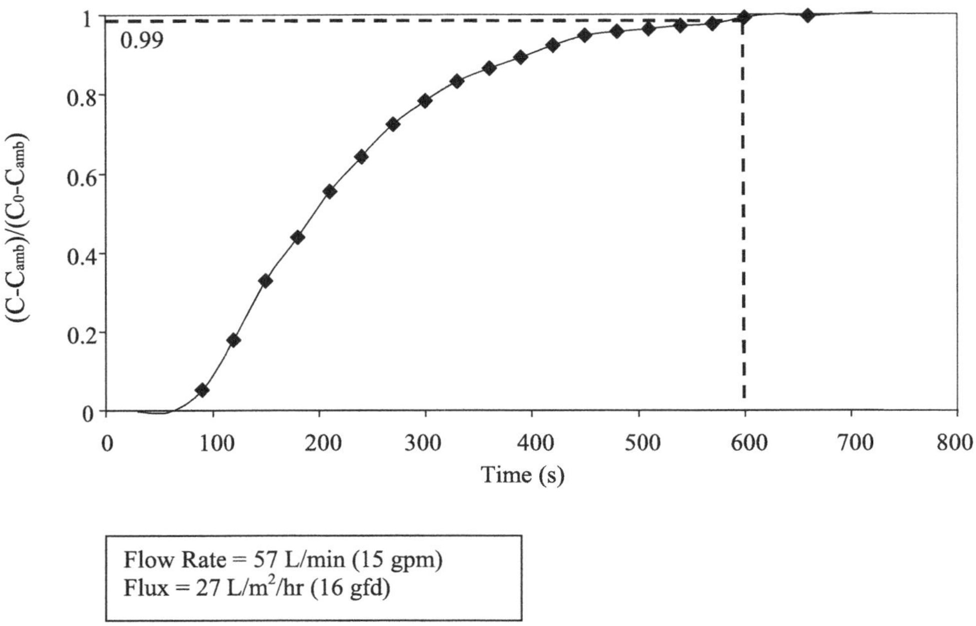

Figure 4.37 Tracer testing results at 27 L/m²/hr (submerged PVDF membrane pilot unit, Altanta)

improved as a function of specific flux decline and remained less than 1.0 log. Removal of PRD1 bacteriophage gradually increased as the specific flux declined. The final additional removal of PRD1 bacteriophage reached 1.6 logs at the end of the two-week testing period.

Experimentation With the Submerged PVDF Membrane Pilot Unit

Hydraulic Characterization. Tracer tests were conducted at three different flow rates on the submerged PVDF membrane pilot unit at the AFC WTP. Figure 4.37 shows the results of the three tracer studies. Figure 4.38 depicts the linear regression of the T_{99} value as a function of flow rate. The number of HRTs to reach the steady-state conditions with the submerged pilot unit ranged from 3.5 to 3.8 (see Figure 4.39). As a conservative approach, it was decided to conduct filtrate sampling after at least 5 HRTs.

Microbial Removal Efficiency Under Clean Water Conditions. Three sets of microbial challenges were conducted with the submerged PVDF membrane pilot unit using MS2 and PRD1 bacteriophage and *P. diminuta* seeded into dechlorinated tap water (finished water from the AFC WTP). For each set of microbial challenges, samples were collected at three different fluxes within the same filtration cycle (no backwash occurred during the test).

Before each microbial challenge, a filtrate sample was collected to verify the absence of microbial contamination. *P. diminuta* were present in the filtrate samples collected prior to the first set of microbial challenges at an average concentration of 63/100mL, which was greater than the actual concentration measured in the filtrate during the first microbial challenge (35 to 40/100mL). Since the media used for the enumeration of *P. diminuta* was non-selective, the presence of other microorganisms interfered with *P. diminuta* enumeration. Additionally, the submerged PVDF membrane pilot unit appeared to be more susceptible to biofilm growth than the

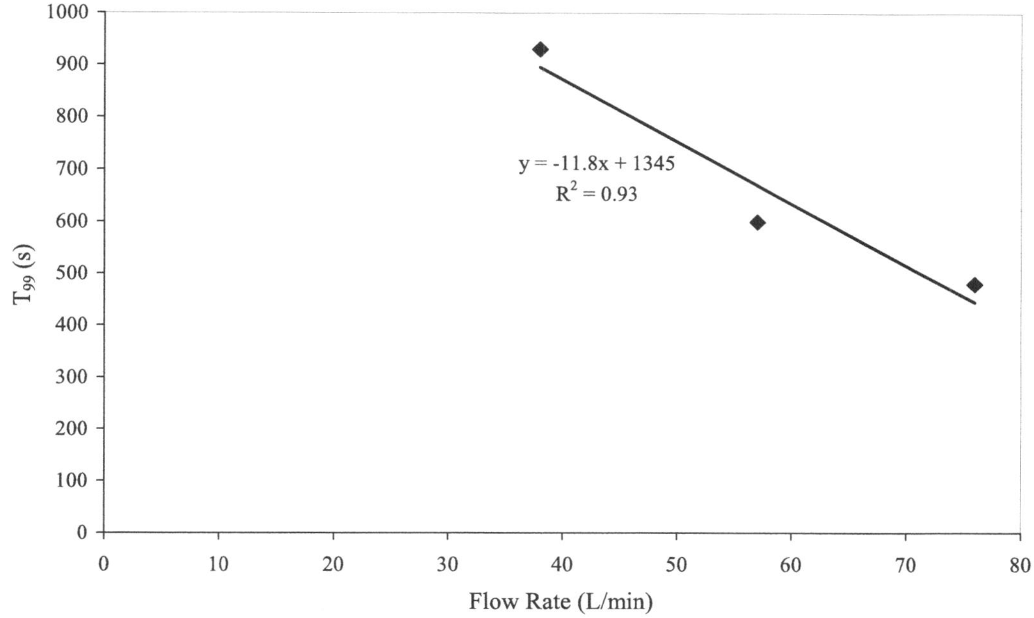

Figure 4.38 T_{99} as a function of flow rate (submerged PVDF membrane pilot unit, Atlanta)

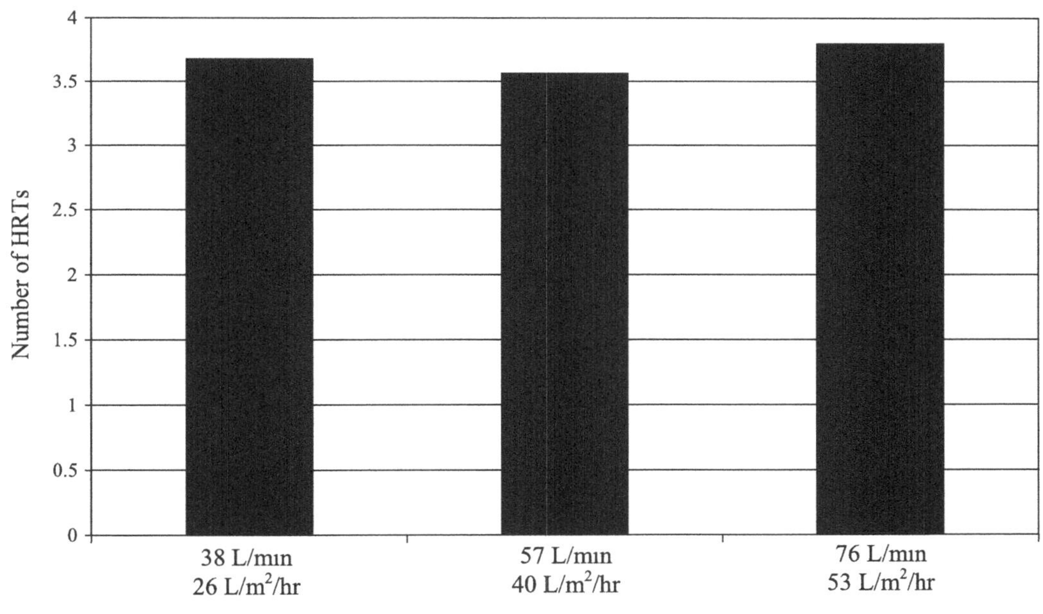

Figure 4.39 Number of HRTs to reach T_{99} (submerged PVDF membrane pilot unit, Atlanta)

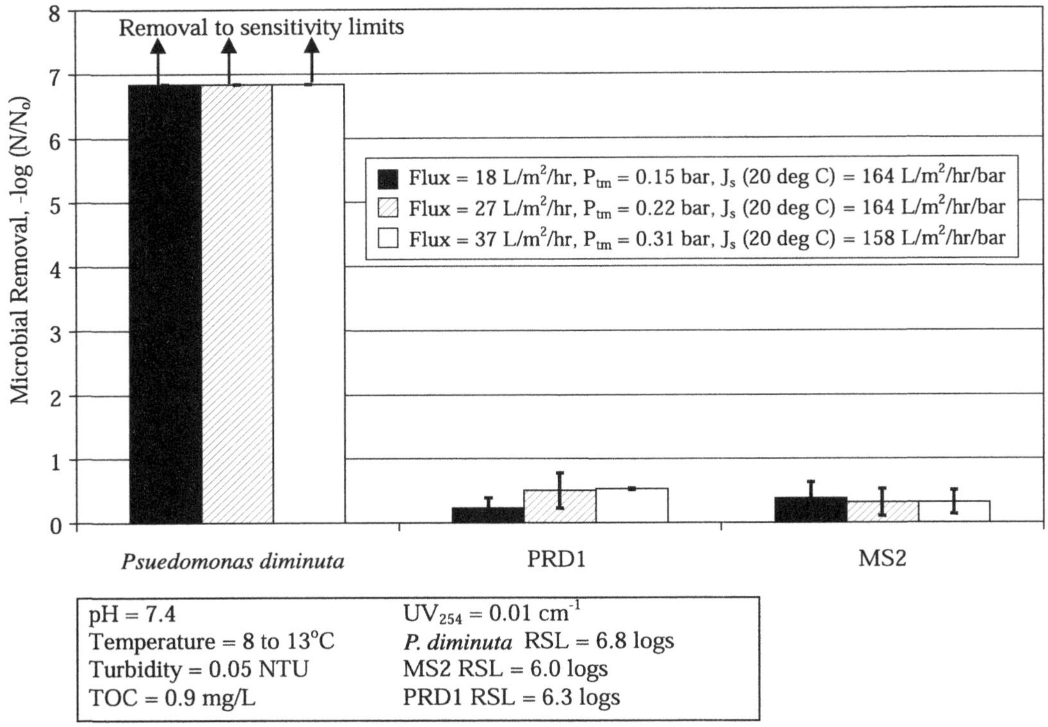

Figure 4.40 Microbial removal under clean water conditions (submerged PVDF membrane pilot unit, Atlanta)

pressurized PVDF membrane pilot unit due to a more complex piping arrangement and a higher potential for dead zones, which made decontamination more difficult. Prior to the third trial, an enhanced decontamination procedure was followed as described in Chapter 2, Methods and Materials, which decreased the HPC level in filtrate samples to acceptable levels prior to the remaining microbial challenge.

Figure 4.40 shows the results obtained for removal of *P. diminuta* and MS2 and PRD1 bacteriophage. For MS2 and PRD1 bacteriophage, error bars were calculated based on the experimental standard deviation of results from triplicate microbial challenges at each condition. Removal to the sensitivity limits of the microbial assay for *P. diminuta* was observed during the third trial after enhanced decontamination procedures were implemented for the submerged PVDF membrane pilot unit.

An additional test was conducted using *Klebsiella terrigena* spiked into dechlorinated tap water. Due to the contamination by HPC bacteria interfering with the enumeration of *P. diminuta* during the first two trials, *K. terrigena* was selected to ascertain the removal to sensitivity limits of bacterial surrogates by the submerged PVDF membrane pilot unit. As shown in Figure 4.41, removal to sensitivity limits of the assay for *K. terrigena* was achieved by the submerged PVDF membrane pilot unit.

Confidence intervals were calculated around the mean for each of the results shown in Figures 4.40 and 4.41 (error bars, not confidence intervals are shown on graph). For *K. terrigena* and MS2 and PRD1 bacteriophage the confidence intervals ranged from 6.4 to 6.7 logs, -0.2 to 0.9 logs, and -0.3 to 0.9 logs respectively. As is clear from the figures, there was a

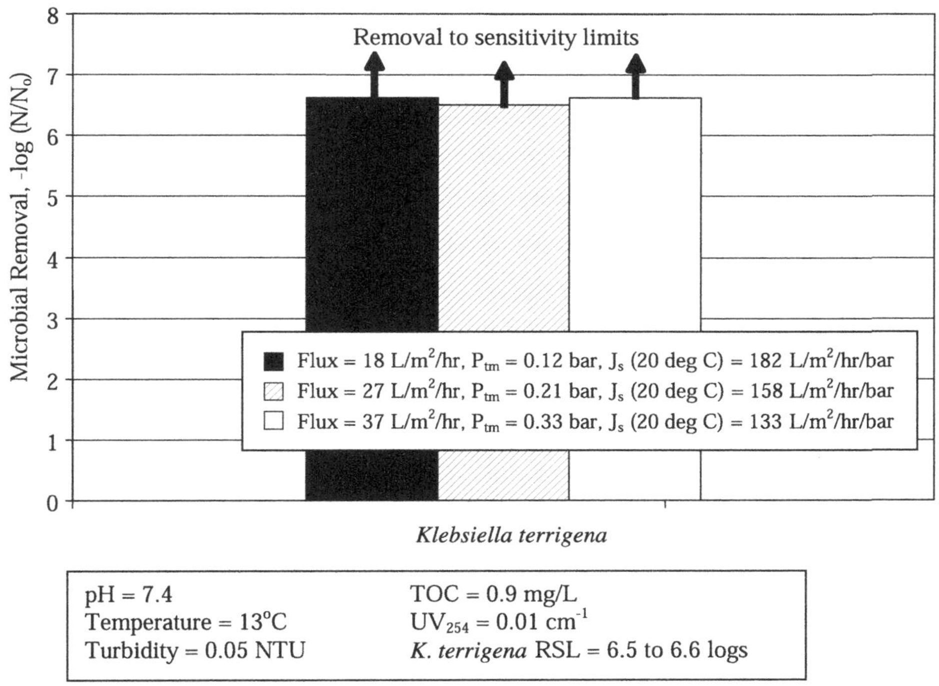

Figure 4.41 *Klebsiella terrigena* removal under clean water conditions (submerged PVDF membrane pilot unit, Atlanta)

statistically significant difference between removal of the bacteria (*P. diminuta* and *K. terrigena*) and the bacteriophage (MS2 and PRD1). There was not a statistically significant difference between the removal of MS2 and PRD1 bacteriophage under clean water conditions by the submerged PVDF membrane pilot unit.

As removal to sensitivity limits of the assay for *P. diminuta* and *K. terrigena* was observed with the submerged PVDF membrane pilot unit during clean water testing, only MS2 and PRD1 bacteriophage were used in subsequent testing with this pilot plant. Removal of MS2 and PRD1 bacteriophage under clean water conditions was used to calculate the net microbial removal for subsequent testing.

Evaluation of the Impact of Deposition Layer Formation on Microbial Removal. While using filtered water with the pressurized PVDF membrane pilot unit, it was observed that deposition layer formation was minimal. Additional microbial removal in filtered water compared to clean water conditions was low as well. Hence, it was decided to test microbial removal in the submerged PVDF membrane pilot unit using only raw water.

Microbial challenge experiments were conducted with the submerged PVDF membrane pilot unit at fluxes of 18, 27, and 37 L/m^2/hr (11, 16, and 22 gfd) using MS2 and PRD1 bacteriophage seeded into raw water. Water quality during these experiments is shown in Table 4.12. Figure 4.42 depicts the net removal of MS2 and PRD1 bacteriophage during the 4-hour filtration cycle for a constant flux of 27 L/m^2/hr (16 gfd). The decline in specific flux during the experiment is also shown. Compared to the pressurized PVDF membrane pilot unit, the decrease in specific flux over the filtration cycle was greater by almost a factor of two. The net removal of PRD1 bacteriophage ranged from 0.9 to 1.8 logs, which was greater than that observed using the

Table 4.12
Raw water quality during deposition layer formation challenges (submerged PVDF membrane pilot unit, Atlanta)

Parameter	Unit	Flow rate 38 L/min Flux 18 L/m²/hr	Flow rate 57 L/min Flux 27 L/m²/hr	Flow rate 76 L/min Flux 37 L/m²/hr
Temperature	°C	14	13	14
Turbidity	NTU	4.2	3.5	3.7
pH		7.0	6.8	6.9
Hardness	mg/L $CaCO_3$	15	15	16
Alkalinity	mg/L $CaCO_3$	15	16	15
TOC	mg/L	1.8	1.4	1.9
TSS	mg/L	<2.3	<2.2	<2.2
UV_{254}	cm^{-1}	0.035	0.039	0.0475

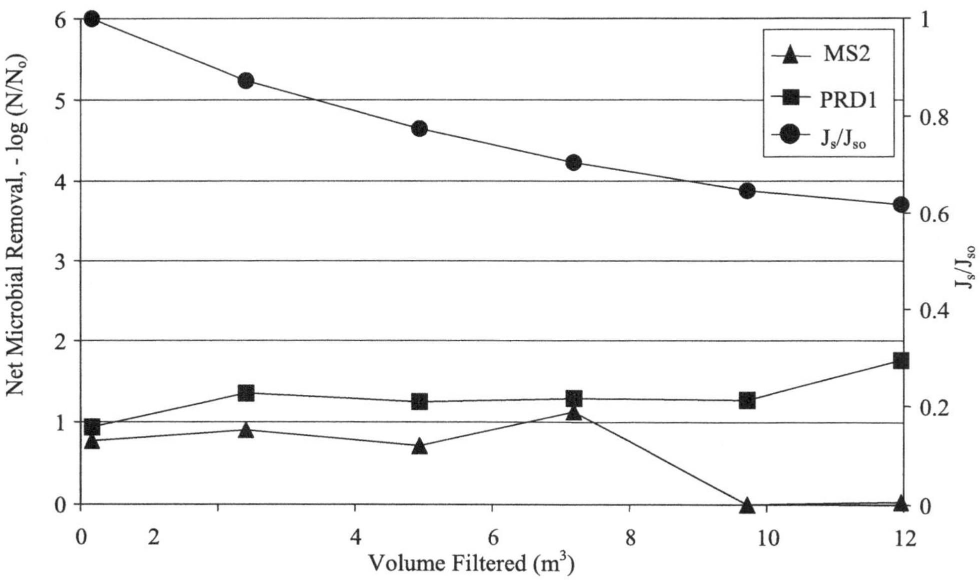

Flux = 27 L/m²/hr (16 gfd)	J_{so} (20°C) = 86 L/m²/hr/bar (3.5 gfd/psi)
Flow Rate = 57 L/min (15 gpm)	MS2 RSL = 6.2 logs
P_{tmo} = 0.38 bar (5.5 psi)	PRD1 RSL = 6.4 logs

Note:
MS2 bacteriophage removal under clean water conditions at 27 L/m²/hr = 0.3 log
PRD1 bacteriophage removal under clean water conditions at 27 L/m²/hr = 0.5 log

Figure 4.42 Impact of deposition layer formation on microbial removal (raw water, submerged PVDF membrane pilot unit, flux = 27 L/m²/hr, Atlanta)

Table 4.13
Additional microbial removal (compared to clean water conditions) observed during a 4-hour filtration cycle (raw water, submerged PVDF membrane pilot unit, Atlanta)

	Flow rate = 38 L/min Flux = 11 L/m²/hr	Flow rate = 57 L/min Flux = 27 L/m²/hr	Flow rate = 76 L/min Flux = 37 L/m²/hr
Total volume filtered	9 m^3	14 m^3	18 m^3
Reduction in specific flux after 4 hours of filtration	28%	38%	33%
Additional removal of MS2 bacteriophage	0.0 log	0.0 to 1.1 logs	0.1 to 0.6 log
Additional removal of PRD1 bacteriophage	0.7 to 1.5 logs	0.9 to 1.8 logs	1.4 to 1.8 logs

pressurized PVDF membrane pilot unit, whereas the additional removal of MS2 bacteriophage varied between 0 and 1.1 logs, which was approximately the same as that observed in the pressurized PVDF membrane pilot unit. Similar net microbial removal was observed at the other fluxes and flow rates tested. Table 4.13 summarizes the additional removal compared to clean water conditions obtained for the three fluxes tested. The higher removal of PRD1 bacteriophage observed in the submerged PVDF membrane pilot unit can be attributed to the greater decrease in specific flux occurring in the submerged PVDF membrane pilot unit compared to the pressurized PVDF membrane pilot unit. Due to natural water quality variations, the raw water used as feedwater for the submerged PVDF membrane pilot unit had slightly higher turbidity and TOC than that used during testing of the pressurized PVDF membrane pilot unit, which may have led to a greater decrease in specific flux during the 4-hour experiment.

Evaluation of the Impact of Membrane Fouling on Microbial Removal. Two experiments were conducted with the submerged PVDF membrane pilot unit. The first experiment was conducted at a flux of 27 L/m^2/hr (16 gfd) with a backwash interval of 15 minutes. The increase in transmembrane pressure occurred more rapidly than anticipated, and the operational limits of the unit were reached after only 30 hours. Two microbial challenges were conducted over this period. The second experiment was performed over 11 days (10.6 days of filtration) at a flux of 18 L/m^2/hr (11 gfd) with a backwash interval of 15 minutes. Five microbial challenges were conducted over this period. Water quality during this experiment is reported in Table 4.14. Figure 4.43 depicts the net removal of MS2 and PRD1 bacteriophage as a function of the filtration time. Specific flux decline is also shown in the figure.

During the first experiment, specific flux decreased by 60% over only 30 hours. During the second experiment, the specific flux declined by 58% over 11 days. Additional MS2 bacteriophage removal compared to clean water conditions was not significantly improved and remained less than 1.0 log in both experiments, regardless of specific flux decline. Additional removal of PRD1 bacteriophage compared to clean water conditions gradually increased as the specific flux declined, ranging from 0.9 to 4.1 logs in the second experiment.

Table 4.14
Raw water quality during membrane fouling challenges
(submerged PVDF membrane pilot unit, Atlanta)

Parameter	Unit	Average (min-max)
Temperature	°C	17 (16-19)
Turbidity	NTU	4.0 (2.4-6.8)
pH		6.9 (6.8-7.1)
Hardness	mg/L $CaCO_3$	17 (16-18)
Alkalinity	mg/L $CaCO_3$	16 (15-18)
TOC	mg/L	1.7 (1.4-1.9)
TSS	mg/L	<2.3
UV_{254}	cm^{-1}	0.042 (0.036-0.047)

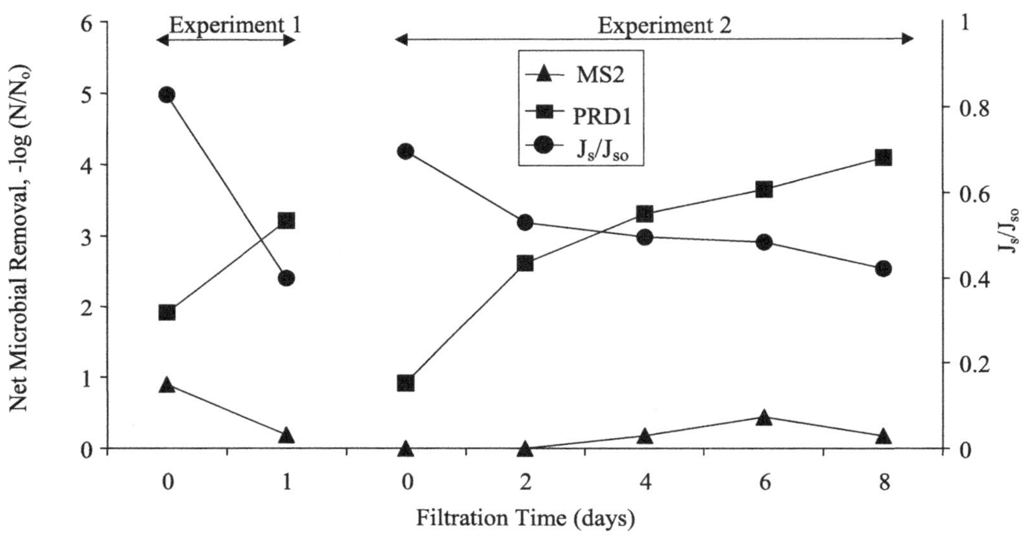

Experiment 1
Flux = 27 L/m^2/hr (16 gfd)
Flow Rate = 57 L/min (15 gpm)
P_{tmo} = 0.31 bar (4.5 psi)
J_{so} (20°C) = 101 L/m^2/hr/bar (4.1 gfd/psi)
Backwash Interval = 15 min
MS2 RSL = 6.1 logs

Experiment 2
Flux = 19 L/m^2/hr (11 gfd)
Flow Rate = 38 L/min (10 gpm)
P_{tmo} = 0.21 bar (3.0 psi)
J_{so} (20°C) = 101 L/m^2/hr/bar (4.1 gfd/psi)
Backwash Interval = 15 min
PRD1 RSL = 5.7 logs

Note:
MS2 bacteriophage removal under clean water conditions at 27 L/m^2/hr = 0.3 log, and at 18 L/m^2/hr = 0.4 log
PRD1 bacteriophage removal under clean water conditions at 27 L/m^2/hr = 0.5 log, and at 18 L/m^2/hr = 0.2 log

Figure 4.43 Impact of fouling on microbial removal over a two-week filtration experiment (raw water, submerged PVDF membrane pilot unit, Atlanta)

Pilot Study at the Scottsdale Water Campus, Arizona

Pilot-scale experimentation at the Scottsdale Water Campus (SWC) consisted of executing Tasks 2 through 4:

- Task 2: Evaluation of microbial removal under clean water conditions
- Task 3: Evaluation of the impact of deposition layer formation on microbial removal
- Task 4: Evaluation of the impact of membrane fouling on microbial removal

Testing was conducted with the pressurized and submerged PVDF membrane pilot units, both of which were equipped with the same 0.1-µm PVDF membrane modules used in Atlanta at the AFC WTP. In Scottsdale, all testing was performed with secondary clarified water from the water reclamation facility.

Experimentation With the Pressurized PVDF Membrane Pilot Unit

Microbial Removal Efficiency Under Clean Water Conditions. Two microbial challenges were conducted with the pressurized PVDF membrane pilot unit using *P. diminuta* and MS2 and PRD1 bacteriophage seeded into dechlorinated tap water. Figure 4.44 depicts the microbial removal achieved at the SWC by the pressurized PVDF membrane pilot unit. Despite aggressive decontamination of the pilot unit with chlorine, heterotrophic plate count (HPC) bacteria were found in the effluent samples at 28 CFU/mL prior to the first microbial challenge, and 167 CFU/mL prior to the second microbial challenge in sampling performed to assess the presence or absence of contamination. During the actual microbial challenges, HPC in the effluent samples remained lower than the concentrations recorded prior to seeding, suggesting that *P. diminuta* was removed up to the sensitivity limit of the assay. No removal of MS2 bacteriophage was observed, while an average removal of 0.3 logs was observed for PRD1 bacteriophage. These values were used to calculate the net microbial removal obtained with the secondary clarified water in subsequent testing of deposition layer formation and fouling using the pressurized PVDF membrane pilot unit. Confidence intervals (95%) were calculated around the mean for each of the results shown in Figure 4.44. For *P. diminuta* and MS2 and PRD1 bacteriophage the confidence limits ranged between 4.6 to 4.7 logs, -0.2 to 0.2 logs, and -0.2 to 1.0 logs respectively. As observed in clean water testing at Atlanta, there was a statistically significant difference between removal of the bacteria and the bacteriophage, but the difference between removal of PRD1 and MS2 bacteriophage was not statistically significant.

Table 4.15 shows a comparison of the results obtained at SWC with those collected at the AFC WTP using the pressurized PVDF membrane pilot unit under clean water conditions. The temperature-corrected specific flux was somewhat similar at AFC WTP and SWC, which demonstrates that the membranes were under similar fouling/maturity conditions at both sites. However, the actual specific flux was much greater at the SWC because of the higher water temperature.

Evaluation of the Impact of Deposition Layer Formation on Microbial Removal. Microbial challenge experiments were conducted with the pressurized PVDF membrane pilot unit at fluxes of 26, 40, and 53 L/m^2/hr (15, 24, and 31 gfd) using MS2 and PRD1 bacteriophage seeded into the wastewater secondary effluent. Water quality recorded during these experiments is shown in Table 4.16.

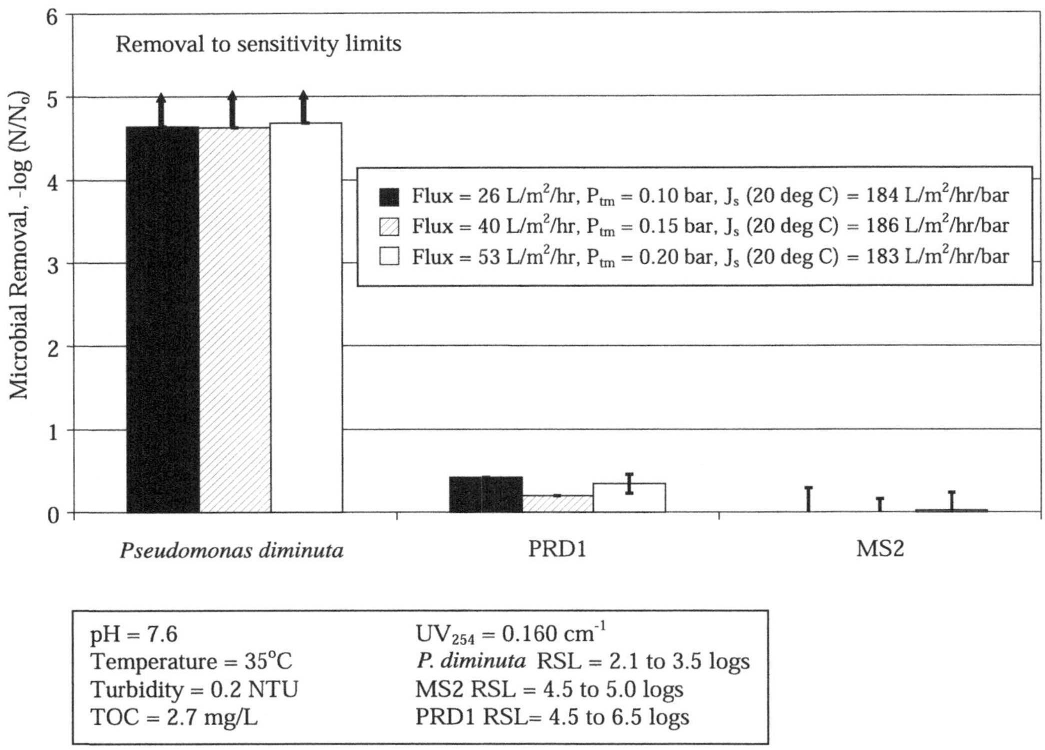

Figure 4.44 Microbial removal under clean water conditions (pressurized PVDF membrane pilot unit, Scottsdale)

Table 4.15
Average microbial removal under clean water conditions at Atlanta and Scottsdale (pressurized PVDF membrane pilot unit)

	Atlanta Fulton County WTP	Scottsdale Water Campus
Specific flux at 20°C	153 to 175 $L/m^2/hr/bar$	177 to 192 $L/m^2/hr/bar$
Actual specific flux	143 to 167 $L/m^2/hr/bar$ at 17°C	246 to 271 $L/m^2/hr/bar$ at 34°C
P. diminuta removal	Up to sensitivity limit (5.1 and 4.8 logs)	Up to sensitivity limit (3.6 and 5.7 logs)
MS2 bacteriophage removal	<0.2 log	<0.2 log
PRD1 bacteriophage removal	1.2 to 1.5 logs	0.3 log

Table 4.16
Secondary effluent quality during deposition layer formation challenges (pressurized PVDF membrane pilot unit, Scottsdale)

Parameter	Unit	Flow rate 38 L/min Flux 26 $L/m^2/hr$	Flow rate 57 L/min Flux 40 $L/m^2/hr$	Flow rate 76 L/min Flux 53 $L/m^2/hr$
Temperature	°C	31	32	33
pH		7.1	7.3	7.2
Turbidity	NTU	2.5	2.9	3.3
TSS	mg/L	<10	<10	<10
TOC	mg/L	5.9	6.3	5.5
UV_{254}	cm^{-1}	0.117	0.116	0.123
COD	mg/L	<20	<20	31
BOD	mg/L	4.7	4.7	2.5
Alkalinity	mg/L $CaCO_3$	202	218	192
Hardness	mg/L $CaCO_3$	294	276	294
Nitrate	mg/L as N	3	2	10
Phosphorus	mg/L	2.6	2.1	2.4
HPC	cfu/mL	3.5×10^4	1.1×10^5	9.1×10^4
Total Coliform	cfu/100mL	1.3×10^6	6.9×10^5	1.3×10^6

Figure 4.45 depicts the net removal of MS2 and PRD1 bacteriophage during the single filtration cycle at a constant flux of 53 $L/m^2/hr$ (31 gfd). The specific flux declined by 50% over 230 minutes of filtration (total volume filtered = 17 m^3). Removal of MS2 bacteriophage was slightly improved during the experiment, with additional removal ranging from 0.8 to 1.6 logs compared to clean water conditions. PRD1 bacteriophage removal was more affected by deposition layer formation and the resulting decline in specific flux, as the additional removal compared to clean water conditions gradually increased during the 4-hour filtration cycle from 1.1 to 3.1 logs. Similar results were observed at the lower fluxes tested (26 and 40 $L/m^2/hr$; 15 and 24 gfd). Table 4.17 summarizes the additional removal recorded at the three fluxes tested compared to clean water conditions. These results demonstrate the impact of deposition layer formation on the removal of PRD1 bacteriophage by the pressurized PVDF membrane pilot unit under the conditions tested when secondary clarified effluent was the membrane feedwater.

Evaluation of the Impact of Membrane Fouling on Microbial Removal. The pressurized PVDF membrane pilot unit was operated for two weeks with the secondary effluent from the SWC at a flux of 40 $L/m^2/hr$ (24 gfd) with a backwash interval of 15 minutes. These operating conditions were set to replicate the conditions tested at the Atlanta Fulton County WTP. Water quality recorded during this experiment is reported in Table 4.18. Six microbial challenge

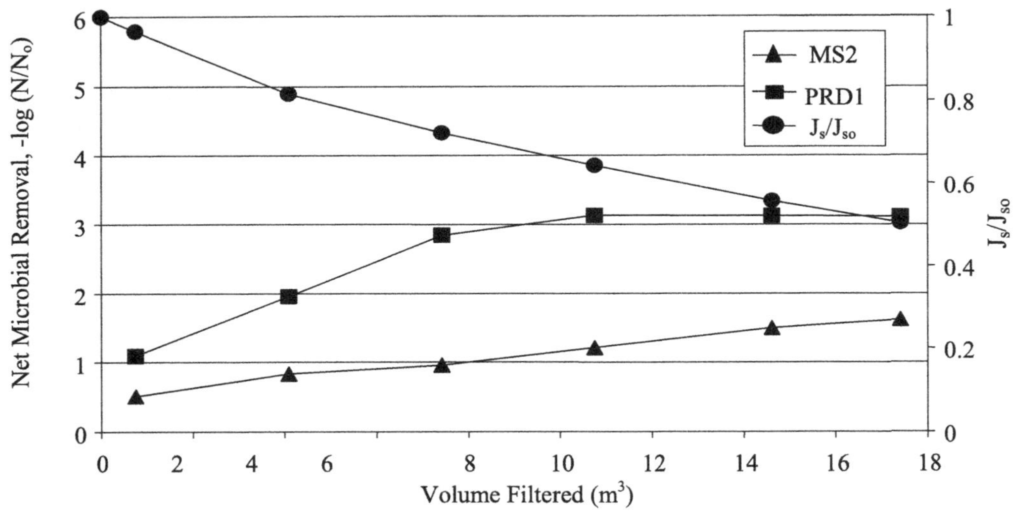

Flux = 53 L/m²/hr (31 gfd)	J_{so} (20°C) = 182 L/m²/hr/bar (7.4 gfd/psi)
Flow Rate = 76 L/min (20 gpm)	MS2 RSL = 4.9 logs
P_{tmo} = 0.21 bar (3.1 psi)	PRD1 RSL = 5.7 logs

Note:
MS2 removal under clean water conditions at 53 L/m²/hr = 0.0 log
PRD1 removal under clean water conditions at 53 L/m²/hr = 0.3 log

Figure 4.45 Impact of deposition layer formation on microbial removal (secondary effluent, pressurized PVDF membrane pilot unit, flux = 53 L/m²/hr, Scottsdale)

Table 4.17
Additional microbial removal (compared to clean water conditions) observed during a 4-hour filtration cycle (secondary effluent, pressurized PVDF membrane pilot unit, Scottsdale)

	Flow rate = 38 L/min Flux = 26 L/m²/hr	Flow rate = 57 L/min Flux = 40 L/m²/hr	Flow rate = 76 L/min Flux= 53 L/m²/hr
Total volume filtered	9 m³	14 m³	18 m³
Reduction in specific flux after 4 hours of filtration	29%	50%	50%
Additional removal of MS2 bacteriophage	0.0 to 0.8 log	0.2 to 0.7 log	0.5 to 1.6 log
Additional removal of PRD1 bacteriophage	0.2 to 2.4 logs	0.7 to 2.6 logs	1.1 to 3.1 logs

Table 4.18
Secondary effluent quality during membrane fouling challenges (pressurized PVDF membrane pilot unit, Scottsdale)

Parameter	Unit	Average	Range
Temperature	°C	33	32-34
pH		7.2	7.1-7.2
Turbidity	NTU	3.4	2.7-3.9
TSS	mg/L	<10	
TOC	mg/L	6.2	6.0-6.6
UV_{254}	cm^{-1}	0.132	0.124-0.143
COD	mg/L	39	35-47
BOD	mg/L	5.7	3.9-8.2
Alkalinity	mg/L $CaCO_3$	200	186-214
Hardness	mg/L $CaCO_3$	307	303-314
TKN	mg/L as N	2.88	1.61-3.99
Nitrate	mg/L as N	3.2	1-5
Phosphorous	mg/L	2.6	2.2-3.2
Heterotrophic Plate Count	CFU/mL	9.9×10^4	7.5×10^4-1.3×10^5
Total Coliform	CFU/100mL	1.7×10^6	1×10^6-2.7×10^6

experiments were conducted over this two-week period (10.6 days of effective filtration) using MS2 and PRD1 bacteriophage. After the second microbial challenge was performed on the third day of operation, a major leak occurred in the pilot unit. The leak was determined to be caused by a loosened PVC pipe collecting the filtrate from the unit's top manifold. The pilot unit was shut down for approximately 30 hours for repair of the PVC and then restarted to resume the experiment under the same conditions.

On the 14th day of operation, immediately after the sixth microbial challenge was performed, the unit was chemically cleaned with citric acid and chlorine. Following this cleaning, a solution of sodium thiosulfate was filtered through the unit in order to quench any chlorine residual, and a seventh microbial challenge was performed in order to assess the impact of chemical cleaning on the removal of MS2 and PRD1 bacteriophage.

Figure 4.46 depicts the decline in specific flux ratio over the two-week testing period. The specific flux constantly decreased as filtration time increased, finally decreasing to 10% of the initial specific flux after 10.6 days of effective filtration. The specific flux after the chemical cleaning was only 33% of the initial specific flux.

Figure 4.47 shows the net removal of MS2 and PRD1 bacteriophage obtained for the seven microbial challenges. Removal of both MS2 and PRD1 bacteriophage increased greatly as the specific flux declined. Additional PRD1 bacteriophage removal due to membrane fouling

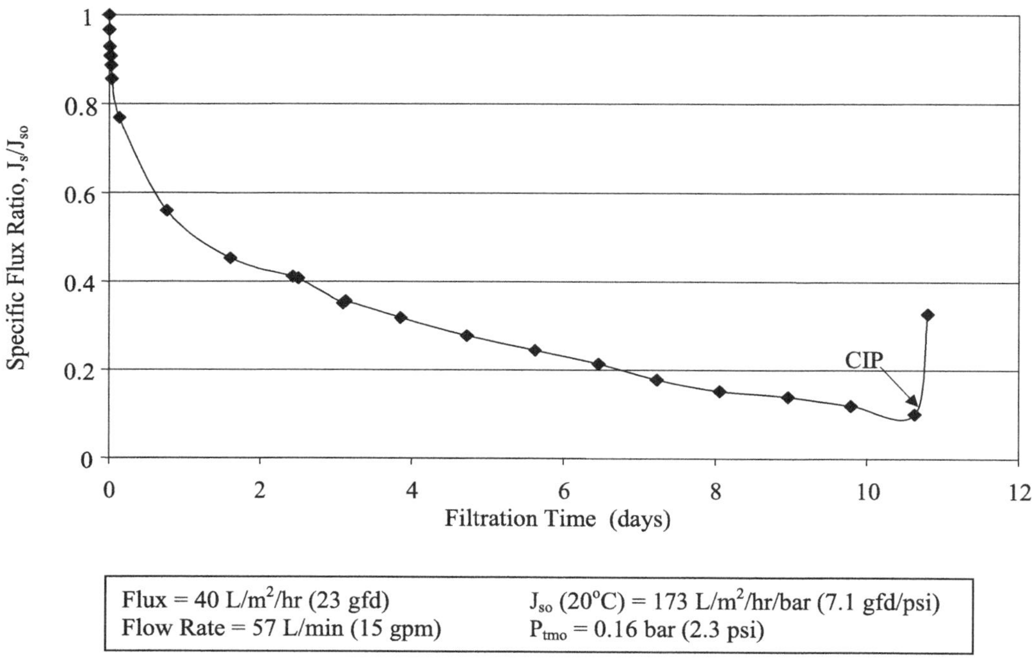

Figure 4.46 Specific flux ratio during membrane fouling challenge testing (secondary effluent, pressurized PVDF membrane pilot unit, Scottsdale)

compared to clean water conditions increased to 5.5 logs after 5 days of filtration and then stabilized at approximately 5.0 logs. Additional MS2 bacteriophage removal compared to clean water conditions reached 4.3 logs at the end of the testing period. The chemical cleaning increased the specific flux from 10% to 33% of the initial flux. The restoration of a portion of the initial flux by the chemical cleaning caused a decrease in bacteriophage removal.

Experimentation With the Submerged PVDF Membrane Pilot Unit

Microbial Removal Efficiency Under Clean Water Conditions. Two tests were conducted on the submerged PVDF membrane pilot unit using *P. diminuta* and MS2 and PRD1 bacteriophage seeded into dechlorinated tap water. Despite aggressive decontamination of the submerged PVDF membrane pilot unit using chlorine, HPC bacteria were found in the background samples (filtrate samples taken prior to microbial seeding) with an average concentration of 25 CFU/mL. During the actual testing, HPC in the effluent samples remained under this background number; thus, *P. diminuta* were removed up to the sensitivity limit of the assay.

Figure 4.48 shows the bacteriophage removal under clean water conditions achieved at the SWC with the submerged PVDF membrane pilot unit. An average removal of 1.3 logs was observed for MS2 bacteriophage, while removal averaged 1.5 logs for PRD1 bacteriophage. These values were used for calculation of the net microbial removal obtained with the secondary clarified water in subsequent microbial challenges evaluating deposition layer formation and membrane fouling in the submerged PVDF membrane pilot unit.

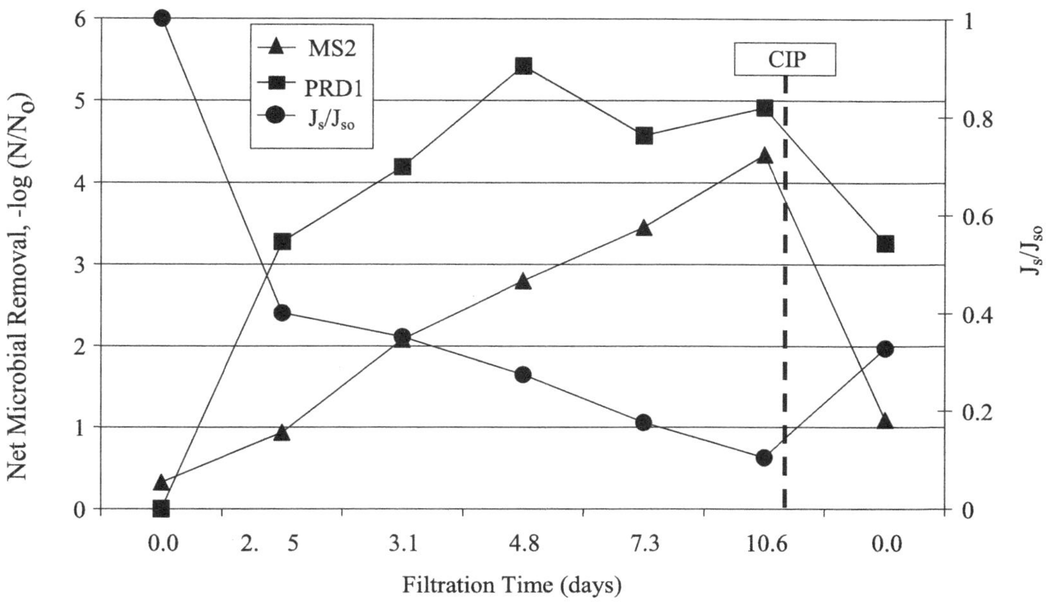

Flux = 40 L/m²/hr (23 gfd)	J_{so} (20°C) = 173 L/m²/hr/bar (7.1 gfd/psi)
Flow Rate = 57 L/min (15 gpm)	MS2 RSL = 5.2 logs
P_{tmo} = 0.16 bar (2.3 psi)	PRD1 RSL = 5.5 logs
Backwash Interval = 15 min	

Note:
MS2 bacteriophage removal under clean water conditions at 40 L/m²/hr = 0.0 log
PRD1 bacteriophage removal under clean water conditions at 40 L/m²/hr = 0.2 log

Figure 4.47 Impact of fouling on microbial removal over a two-week experiment (secondary effluent, pressurized PVDF membrane pilot unit, flux = 40 L/m²/hr, Scottsdale)

Confidence intervals (95%) were calculated around the mean for each of the results shown in Figure 4.48. For *P. diminuta* and MS2 and PRD1 bacteriophage, confidence limits ranged from 2.6 to 5.3 logs, 0.4 to 2.3 logs and 0.1 to 2.9 logs, respectively.

Table 4.19 provides a comparison with the previous results obtained at the AFC WTP with the submerged PVDF membrane pilot unit under clean water conditions. When the submerged PVDF membrane pilot unit was tested under clean water conditions at the AFC WTP, the PVDF membrane modules were brand new, which resulted in a higher specific flux compared to that observed at the SWC. The increased level of maturity of the modules observed at SWC resulted not only in a lower specific flux, but also in a discernibly higher bacteriophage removal. The inability to chemically clean the membranes to their original specific flux suggests that these membranes may have been irreversibly fouled to some extent.

Evaluation of the Impact of Deposition Layer Formation on Microbial Removal. Microbial challenge experiments were conducted with the submerged PVDF membrane pilot unit at fluxes of 18, 27, and 37 L/m²/hr (11, 16, and 22 gfd) using MS2 and PRD1 bacteriophage seeded into SWC secondary effluent. Water quality during these experiments is provided in Table 4.20.

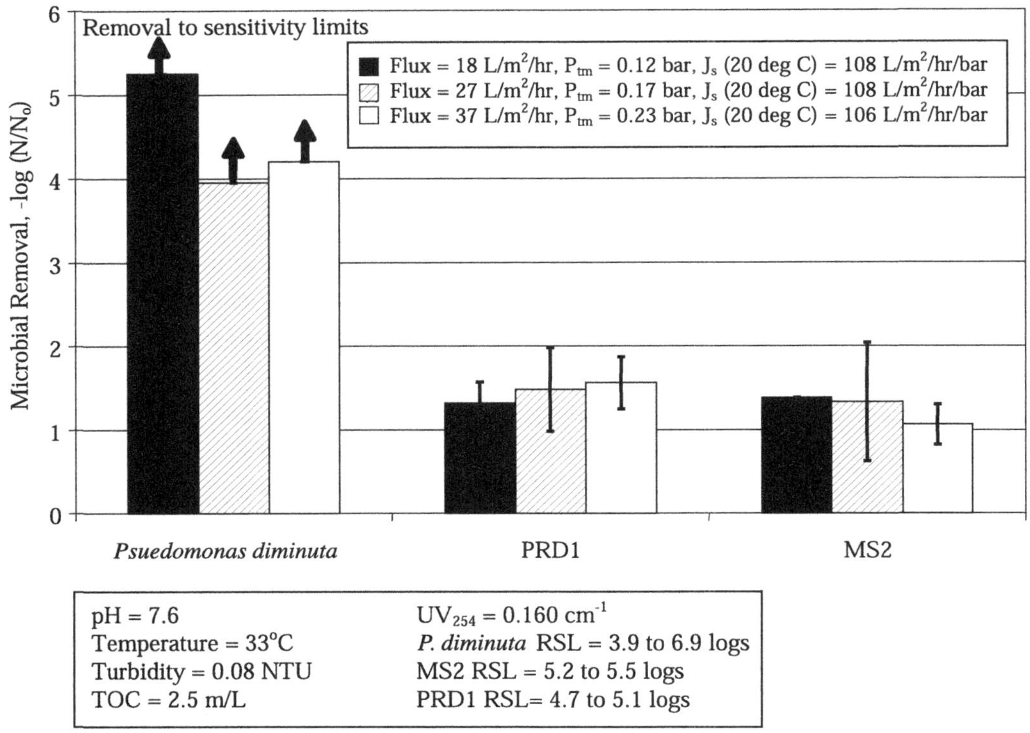

Figure 4.48 Microbial removal under clean water conditions (submerged PVDF membrane pilot unit, Scottsdale)

Table 4.19
Microbial removal under clean water conditions at Atlanta and Scottsdale (submerged PVDF membrane pilot unit)

	AFC WTP	SWC
Specific flux at 20°C	158 to 180 L/m^2/hr/bar	106 to 108 L/m^2/hr/bar
Actual specific flux	116 to 123 L/m^2/hr/bar at 8 to 13°C	150 to 161 L/m^2/hr/bar at 33°C
MS2 bacteriophage removal	0.3 log (0.2 to 0.5)	1.3 logs (1.0 to 1.5)
PRD1 bacteriophage removal	0.5 log (0.3 to 0.5)	1.5 logs (1.0 to 1.5)

As shown in Table 4.21, the decreases in specific flux occurring in the submerged PVDF membrane pilot unit were greater than those observed in the pressurized PVDF membrane pilot unit. Similar observations were made during the testing at the Atlanta Fulton County WTP.

Figure 4.49 depicts the net removal of MS2 and PRD1 bacteriophage during a single filtration cycle for a constant flux of 27 L/m^2/hr (16 gfd). The decline in specific flux during this experiment is also shown. The additional removal of MS2 bacteriophage compared to clean water conditions ranged from 0.4 to 1.3 logs. The additional removal of PRD1 bacteriophage compared

Table 4.20
Secondary effluent quality during deposition layer formation challenges (submerged PVDF membrane pilot unit, Scottsdale)

Parameter	Unit	Flow rate 38 L/min Flux 11 L/m²/hr	Flow rate 57 L/min Flux 16 L/m²/hr	Flow rate 76 L/min Flux 21 L/m²/hr
Temperature	°C	30	30	30
pH		7.1	7.1	7.1
Turbidity	NTU	2.2	2.2	2.2
TSS	mg/L	<10	<10	<10
TOC	mg/L	6.0	5.9	5.8
UV_{254}	cm^{-1}	0.106	0.110	0.104
COD	mg/L	21	23	19
Alkalinity	mg/L $CaCO_3$	188	184	189
Hardness	mg/L $CaCO_3$	323	298	303
Nitrate	mg/L as N	3	5	5
Phosphorus	mg/L	2.4	2.3	2.2
HPC	cfu/mL	8.0×10^4	4.0×10^4	6.0×10^4
Total Coliform	cfu/100mL	7.5×10^5	6.3×10^5	8.8×10^5

Table 4.21
Additional removal (compared to clean water conditions) observed during a 4-hour filtration cycle (secondary effluent, submerged PVDF membrane pilot unit, Scottsdale)

	Flow rate = 38 L/min Flux = 18 L/m²/hr	Flow rate = 57 L/min Flux = 27 L/m²/hr	Flow rate = 76 L/min Flux = 37 L/m²/hr
Total volume filtered	9 m^3	14 m^3	18 m^3
Reduction in specific flux after 4 hours of filtration	55%	53%	61%
Additional removal of MS2 bacteriophage	0.4 to 1.3 logs	0.6 to 0.9 log	0.7 to 1.0 log
Additional removal of PRD1 bacteriophage	1.1 to 2.0 logs	0.7 to 1.5 logs	1.3 to 1.9 logs

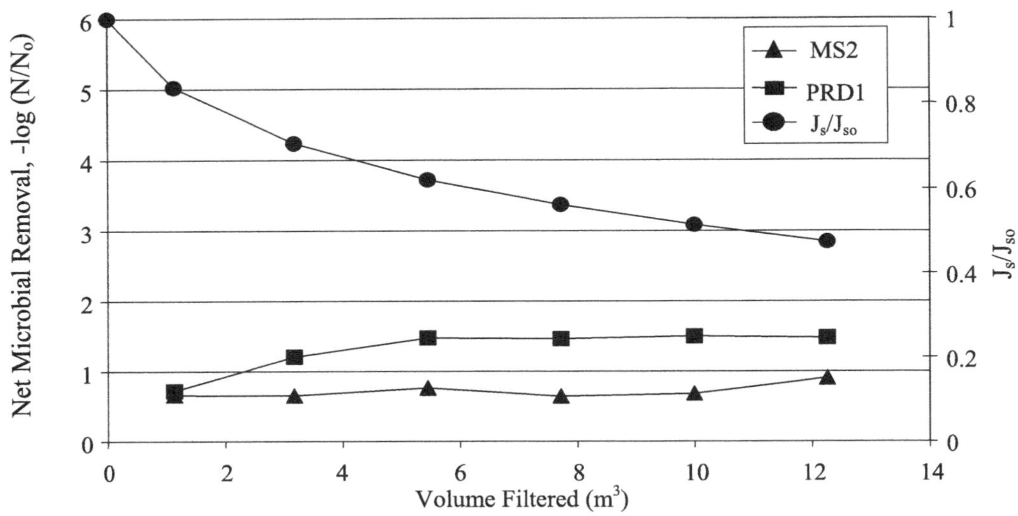

Flux = 27 L/m²/hr (16 gfd)	J_{so} (20°C) = 88 L/m²/hr/bar (3.6 gfd/psi)
Flow Rate = 57 L/min (15 gpm)	MS2 RSL = 5.3 logs
P_{tmo} = 0.25 bar (3.6 psi)	PRD1 RSL = 5.3 logs

Note:
MS2 bacteriophage removal under clean water conditions at 27 L/m²/hr = 1.3 logs
PRD1 bacteriophage removal under clean water conditions at 27 L/m²/hr = 1.5 logs

Figure 4.49 Impact of deposition layer formation on microbial removal (secondary effluent, submerged PVDF membrane pilot unit, flux = 27 L/m²/hr, Scottsdale)

to clean water conditions varied between 0.7 and 1.5 logs. Similar results were achieved at the two other fluxes tested. Table 4.21 summarizes the additional microbial removal compared to clean water conditions observed for the three fluxes tested. Despite the relatively large decreases in specific flux over the four-hour experiments, very little influence on bacteriophage removal was observed at the three fluxes.

Evaluation of the Impact of Membrane Fouling on Microbial Removal. The submerged PVDF membrane pilot unit was operated for two weeks at a flux of 18 L/m²/hr (11 gfd) with a backwash interval of 15 min. Six microbial challenges were conducted over this two-week period. After the first microbial challenge was performed on the first day of operation, the pilot feed pump failed, forcing the unit to be shut down for two days for repairs before resuming the experiment under the same conditions.

Figure 4.50 depicts the specific flux decline over the testing period. After the sixth day (two days of filtration, followed by two days of shutdown, followed by two days of filtration), the specific flux stabilized at 30% of the initial specific flux and remained very stable until the end of the testing period. The flux decline over the two weeks of testing in the submerged PVDF membrane pilot unit was less than that observed for the pressurized PVDF membrane pilot unit. This difference could be due to lower turbidity, COD and microorganisms (HPC, total coliform) in the secondary effluent fed to the submerged PVDF membrane pilot unit compared to the feed-water for testing with the pressurized PVDF membrane pilot unit, or to a greater efficiency of backwashing.

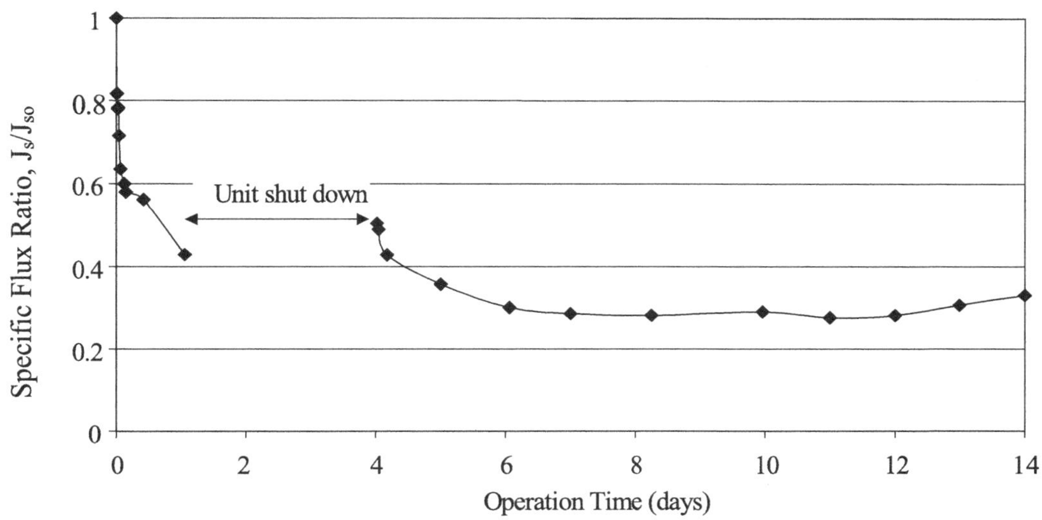

Flux = 18 L/m^2/hr (11 gfd)
Flow Rate = 38 L/min (10 gpm)
Backwash Interval = 30 min
P_{tmo} = 0.12 bar (1.8 psi)
J_{so} (20°C) = 112 L/m^2/hr/bar (4.5 gfd/psi)

Figure 4.50 Specific flux ratio during membrane fouling challenge testing (secondary effluent, submerged PVDF membrane pilot unit, Scottsdale)

Figure 4.51 shows the net removal of MS2 and PRD1 bacteriophage as a function of the filtration time. Additional removal of PRD1 bacteriophage compared to clean water conditions ranged from 1.3 to 3.1 logs and exceeded 2.6 logs after the specific flux stabilized. No additional removal of MS2 bacteriophage compared to clean water conditions was observed.

PILOT-SCALE STUDIES - UNITED KINGDOM

Pilot testing in the UK followed the same protocol as that in Atlanta and Scottsdale. Testing was performed using two membrane pilot systems: one submerged system and one pressurized system. The protocol included the following tasks:

- Task 1: Hydraulic characterization
- Task 2: Evaluation of microbial removal under clean water conditions
- Task 3: Evaluation of the impact of deposition layer formation on microbial removal
- Task 4: Evaluation of the impact of membrane fouling on microbial removal

These tasks were performed at the Kempton Park Advanced Water Treatment Works (AWTW) of Thames Water Utilities in the London area. The tasks were performed using two different membrane pilot units: a submerged PVDF membrane pilot unit, designated "PVDF" and a pressurized PES/polyvinyl pyrrolidone (PVP) membrane pilot unit, designated "PES/PVP." The raw water was from the Queen Mary Reservoir, fed by the River Thames. Filtered water for this supply was treated by slow sand filtration followed by chloramine disinfection, with dechlorination prior to use as feedwater to the membrane pilot units.

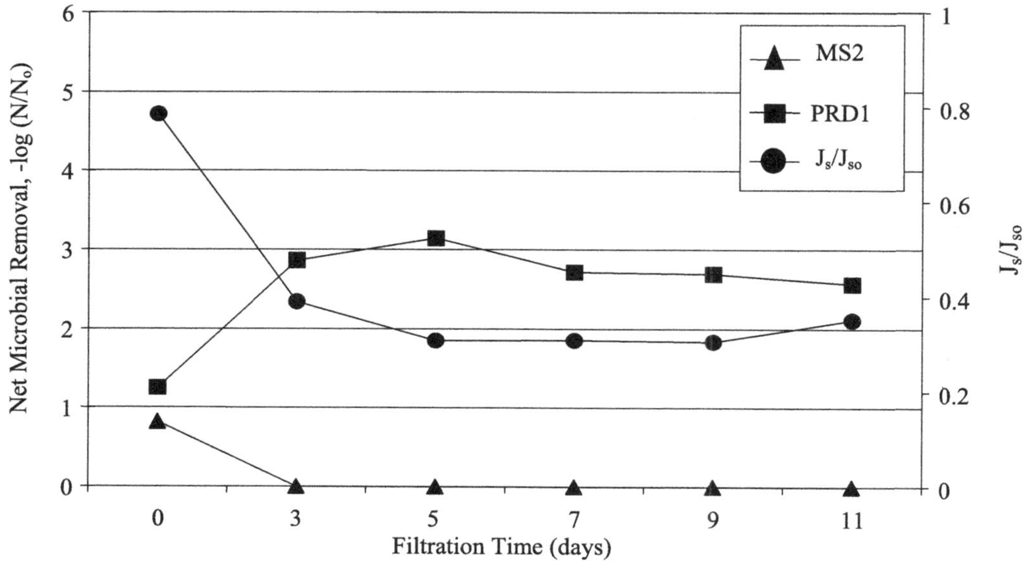

Flux = 18 L/m^2/hr (11 gfd)
Flow Rate = 38 L/min (10 gpm)
P_{tmo} = 0.12 bar (1.8 psi)
Backwash Interval = 30 min
J_{so} (20°C) = 112 L/m^2/hr/bar (4.5 gfd/psi)
MS2 RSL = 4.6 log
PRD1 RSL = 5.8 log

Note:
MS2 bacteriophage removal under clean water conditions at 18 L/m^2/hr = 1.4 logs
PRD1 bacteriophage removal under clean water conditions at 18 L/m^2/hr = 1.3 logs

Figure 4.51 Impact of fouling on microbial removal over a two-week filtration experiment (secondary effluent, submerged PVDF membrane pilot unit, flux = 18 L/m^2/hr, Scottsdale)

Experimentation With the Submerged PVDF Membrane Pilot Unit

Hydraulic Characterization

The residence time distribution of the submerged PVDF membrane pilot unit was determined by conducting tracer tests at three different fluxes (49, 73 and 86 L/m^2/hr; 29, 43 and 51 gfd). Figure 4.52 shows the results obtained from tracer testing conducted at 73 L/m^2/hr (43 gfd). Figure 4.53 depicts a linear regression of T_{99} as a function of flow rate. T_{99} represents the time required for 99% of the incoming tracer concentration to be recorded in the product stream. The number of HRTs to reach a steady state condition was calculated from the tracer tests. Figure 4.54 shows the number of HRTs required to achieve steady-state conditions for the three fluxes tested.

Microbial Removal Efficiency Under Clean Water Conditions

Microbial challenges were performed on the submerged PVDF membrane pilot unit with dechlorinated tap water at three flux rates during the same filtration cycle. Chlorine residual was quenched with the addition of sodium thiosulfate, and zero residual was confirmed by on-site testing. Microorganisms were added directly to the feed tank with the inflow off, as described in

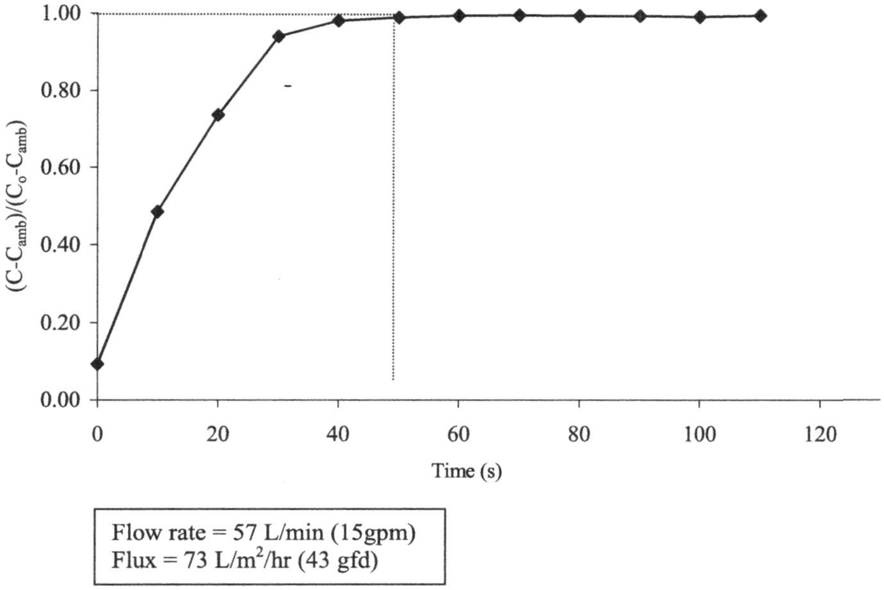

Figure 4.52 Tracer testing results with the 73 L/m²/hr (submerged PVDF membrane pilot unit, London)

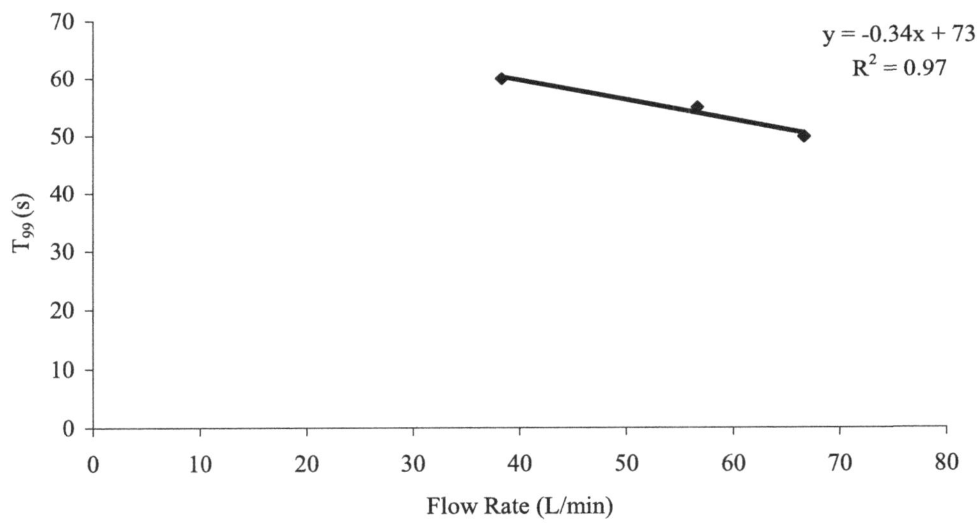

Figure 4.53 T_{99} as a function of flow rate (submerged PVDF membrane pilot unit, London)

Chapter 2, Methods and Materials. The filtration cycle was initiated at 50 L/m²/hr (29 gfd) then increased to 73 L/m²/hr (43 gfd) and finally to 86 L/m²/hr (51 gfd). Influent and effluent samples were collected in duplicate for each flux.

Results are plotted in Figure 4.55. The data indicate complete removal of *P. diminuta*, a 2- to 3-log removal of PRD1 bacteriophage and a 2- to 3.5-log removal of MS2 bacteriophage. *P. diminuta* and PRD1 bacteriophage were removed to the sensitivity limits of the microbial assay. Error bars for MS2 bacteriophage represent the standard deviation between duplicate experiments.

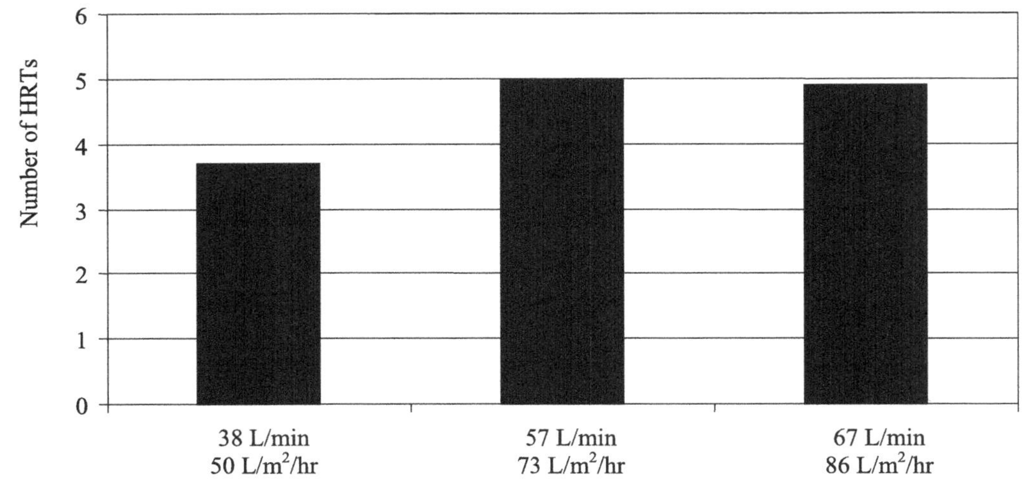

Figure 4.54 Number of HRTs to reach T_{99} (submerged PVDF membrane pilot unit, London)

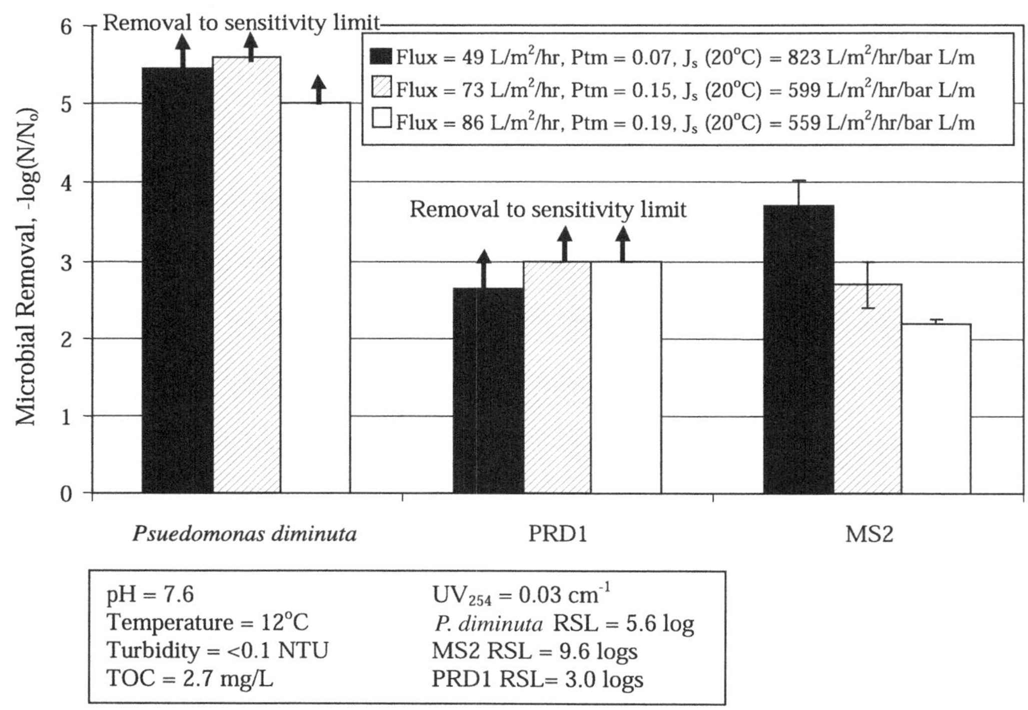

Figure 4.55 Microbial removal under clean water conditions (submerged PVDF membrane pilot unit, London)

Table 4.22
Raw water quality during deposition layer formation challenges (submerged PVDF membrane pilot unit, London)

Parameter	Unit	Flow rate 38 L/min Flux 49 L/m²/hr	Flow rate 57 L/min Flux 73 L/m²/hr	Flow rate 76 L/min Flux 85 L/m²/hr
Temperature	°C	15	17	16
Turbidity	NTU	0.71	0.56	0.42
pH		8.5	8.4	8.4
Hardness	mg/L $CaCO_3$	269	277	292
Alkalinity	mg/L $CaCO_3$	202	209	221
TOC	mg/L	5.0	5.0	5.2
TSS	mg/L	3.5	5.0	<2.0
UV_{254}	cm^{-1}	0.09	0.09	0.09

The results of the clean water experiments were used as the baseline removal for subsequent tests in which the impact of deposition layer formation and membrane fouling on microbial removal was examined in the submerged PVDF membrane pilot unit. Complete removal of *P. diminuta* was expected based on the size of this organism compared to the manufacturer-reported membrane pore size or MWCO. As a result, subsequent experiments were performed with MS2 and PRD1 bacteriophage only. The sensitivity of the PRD1 bacteriophage assay performed for the clean water testing with the submerged PVDF membrane pilot unit was lower than that of other testing in the US and the UK. The lower PRD1 bacteriophage assay sensitivity limit was partially due to the high dilution factor when PRD1 bacteriophage stock suspension was added to 2 m^3 of dechlorinated tap water and the relatively low titer of bacteriophage produced in the laboratory (10^7 to 10^8 pfu/mL). High levels of MS2 bacteriophage stock (10^{12} to 10^{13} pfu/mL) enabled higher levels of sensitivity to be achieved. In subsequent experiments, lower levels of MS2 bacteriophage were employed in microbial challenge tests.

Evaluation of the Impact of Deposition Layer Formation on Microbial Removal

Raw Water

Microbial challenges using MS2 and PRD1 bacteriophage were performed at three different fluxes with the submerged PVDF membrane pilot unit as described in Chapter 2, Methods and Materials. Six microbial challenges were performed over a 4-hour period with no backwash. Water quality parameters associated with each experiment are shown in Table 4.22. PRD1 bacteriophage were removed to the sensitivity limit of the assay. Compared to MS2 bacteriophage removal under clean water conditions, additional removal of MS2 bacteriophage was variable, ranging from 0.0 to 0.9 logs as shown in Figure 4.56 for a flux of 73 L/m²/hr (43 gfd). Negligible additional removal of MS2 bacteriophage was also observed at fluxes of 49 L/m²/hr (29 gfd) and 85 L/m²/hr (51 gfd). Table 4.23 summarizes the results obtained and the decline in specific flux observed.

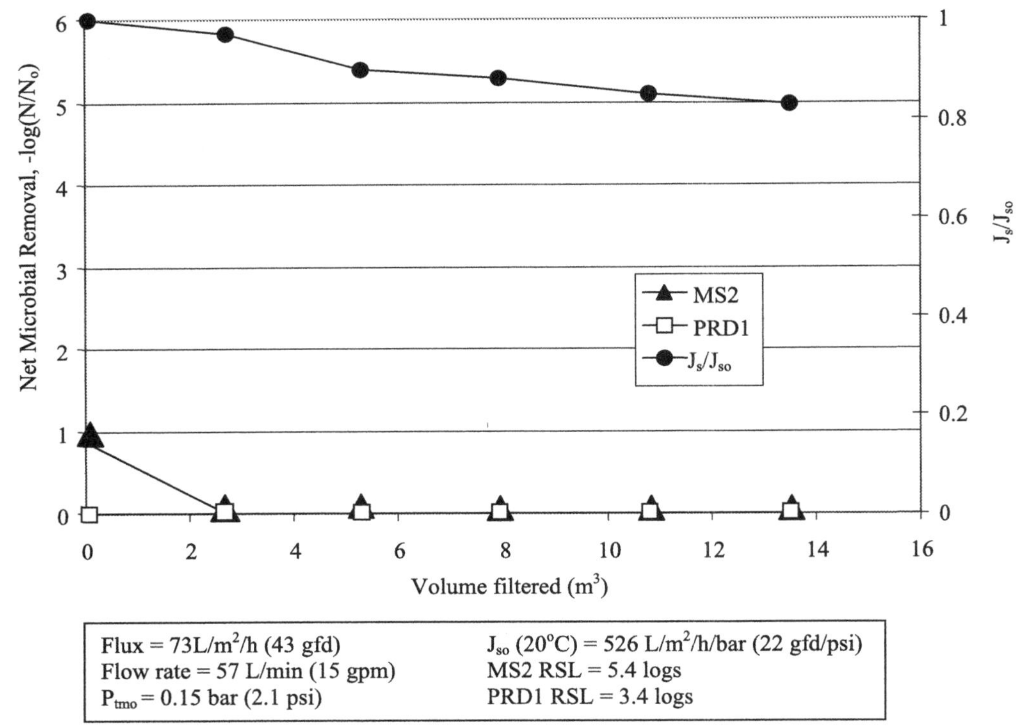

Note:
MS2 bacteriophage removal under clean water conditions at 73 L/m^2/hr = 2.5 logs
PRD1 bacteriophage removal under clean water conditions at 73 L/m^2/hr was at the sensitivity limit of the assay (3.0 logs)

Figure 4.56 Impact of deposition layer formation on microbial removal (raw water, submerged PVDF membrane pilot unit, flux = 73 L/m^2/h, London)

Filtered Water

Microbial challenges were conducted using the filtered water as the feedwater at three different fluxes with MS2 and PRD1 bacteriophage seeded into the filtration tank of the submerged PVDF membrane pilot unit. Average water quality during these experiments is shown in Table 4.24. As shown in Figure 4.57, additional removal compared to clean water conditions was negligible over the 4-hour test for both MS2 and PRD1 bacteriophage at fluxes of 49 L/m^2/hr (29 gfd) and 73 L/m^2/hr (43 gfd). Additional removal of MS2 bacteriophage compared to clean water conditions at a flux of 85 L/m^2/hr (51 gfd) ranged from 0.5 to 1.9 logs, as reported in Table 4.25.

Evaluation of the Impact of Membrane Fouling on Microbial Removal

Raw Water

Experiments were performed to evaluate the impact of fouling on microbial removal using the submerged PVDF membrane pilot unit. Two raw water trials were performed for 4 days each before the plant automatically shut down due to reaching the maximum transmembrane pressure. Water quality measured during the membrane fouling testing is shown in Table 4.26.

Table 4.23
Additional microbial removal (compared to clean water) conditions observed during a 4-hour filtration cycle (raw water, submerged PVDF membrane pilot unit, London)

	Flow rate 38 L/min Flux 48 $L/m^2/hr$	Flow rate 57 L/min Flux 73 $L/m^2/hr$	Flow rate 76 L/min Flux 86 $L/m^2/hr$
Total volume filtered	9 m^3	14 m^3	18 m^3
Reduction in specific flux after 4 hours of filtration	<1%	25%	26%
Additional removal of MS2 bacteriophage	<0.1 log	<0.1 to 0.9 log	<0.1 to 0.9 log
Additional removal of PRD1 bacteriophage	<0.1 log	<0.1 log	<0.1 log

Table 4.24
Filtered water quality during deposition layer formation challenges (submerged PVDF membrane pilot unit, London)

Parameter	Unit	Flow rate 38 L/min Flux 49 $L/m^2/hr$	Flow rate 57 L/min Flux 73 $L/m^2/hr$	Flow rate 76 L/min Flux 86 $L/m^2/hr$
Temperature	°C	18	18	18
Turbidity	NTU	0.1	0.1	0.1
pH		7.6	7.6	7.6
Hardness	mg/L $CaCO_3$	269	285	275
Alkalinity	mg/L $CaCO_3$	209	223	208
TOC	mg/L	2.0	2.4	2.9
TSS	mg/L	2.5	2.0	2.5
UV_{254}	cm^{-1}	0.031	0.028	0.028

The first challenge indicated that microbial removal was variable but higher than that observed under clean water conditions. As shown in Figure 4.58, MS2 bacteriophage removal ranged from 0.5 to 1.5 logs. A second challenge was performed, with results illustrated in Figure 4.59. During the second challenge, the specific flux declined by over 50% and the transmembrane pressure increased from 0.09 to 0.18 bar (1.2 to 2.6 psi). However, as shown in Figure 4.59, additional removal of MS2 bacteriophage was minimal compared to clean water conditions.

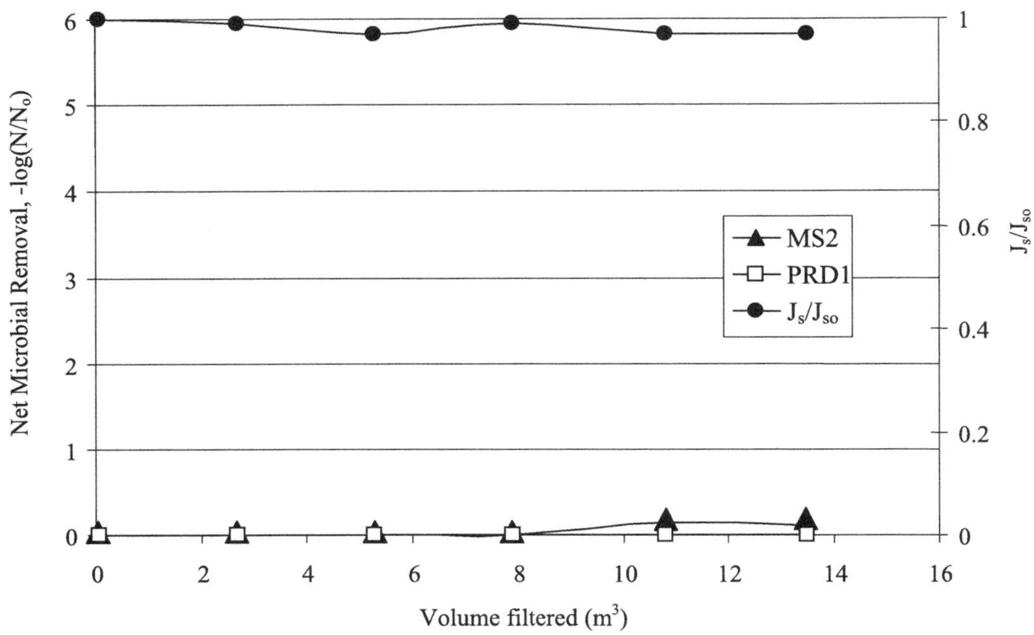

Flux = 73 L/m^2/h (43 gfd)	J_{so} (20°C) = 526 L/m^2/h/bar (22 gfd/psi)
Flow rate = 57 L/min (15 gpm)	MS2 RSL = 5.4 logs
P_{tmo} = 0.15 bar (2.1 psi)	PRD1 RSL= 3.4 logs

Note:
MS2 bacteriophage removal under clean water conditions at 73 L/m^2/hr = 2.5 logs
PRD1 bacteriophage removal under clean water conditions at 73 L/m^2/hr was at the sensitivity limit of the assay (3.0 logs)

Figure 4.57 Impact of deposition layer formation on microbial removal (filtered water, submerged PVDF membrane pilot unit, flux = 73 L/m^2/hr, London)

Table 4.25
Additional microbial removal (compared to clean water conditions) observed during a 4-hour filtration cycle (filtered water, submerged PVDF membrane pilot unit, London)

	Flow rate 38 L/min Flux 49 L/m^2/hr	Flow rate 57 L/min Flux 73 L/m^2/hr	Flow rate 76 L/min Flux 86 L/m^2/hr
Total volume filtered	9 m^3	14 m^3	18 m^3
Reduction in specific flux after 4 hours of filtration	2%	3%	8%
Additional removal of MS2 bacteriophage	<0.1 log	<0.1 log	0.5 to 1.9 logs
Additional removal of PRD1 bacteriophage	<0.1 log	<0.1 log	<0.1 log

Table 4.26
Raw water quality during membrane fouling challenges (submerged PVDF membrane pilot unit, London)

Parameter	Unit	Average (min-max) trial 1	Average (min-max) trial 2
Temperature	°C	19 (18-19)	20 (19-21)
Turbidity	NTU	2.0 (0.92-4.1)	1.1 (0.82-1.3)
pH		8.2 (8.1-8.5)	8.0 (7.9-8.1)
Hardness	mg/L $CaCO_3$	281 (265-294)	293 (289-297)
Alkalinity	mg/L $CaCO_3$	215 (210-224)	215 (211-220)
TOC	mg/L	5.0 (4.7-5.1)	5.0 (4.4-5.3)
TSS	mg/L	3.5 (2.5-4.5)	3.8 (<2-6.5)
UV_{254}	cm^{-1}	0.051 (0.01-0.09)	0.051 (0.01-0.08)

Filtered Water

Testing of filtered water was conducted in the submerged PVDF membrane pilot unit over an 8-day period at a flux of 73 L/m^2/hr (43 gfd). Six microbial challenges were performed during that time. Removal of PRD1 bacteriophage up to the sensitivity limit of the assay (3.4 logs) was observed, as was the case under clean water conditions. Additional removal of MS2 bacteriophage compared to clean water conditions ranged from 0.1 to 1.3 logs. Table 4.27 provides the average and range of water quality observed. Figure 4.60 illustrates the additional removal of MS2 bacteriophage compared to clean water conditions over the 8-day period as well as the observed reduction in specific flux.

Experimentation With the Pressurized PES/PVP Membrane Pilot Unit

Hydraulic Characterization

The residence time distribution of the pressurized PES/PVP pilot unit was determined by conducting tracer tests at three different fluxes (43, 70 and 96 L/ m^2/hr (25, 41 and 57 gfd, respectively). Figure 4.61 shows the results obtained for a flux of 70 L/m^2/hr (41 gfd). Figure 4.62 shows the linear regression of T_{99} as a function of flow rate. Results indicate that the number of HRTs required to achieve steady-state conditions ranged from 2 to 4 for fluxes between 43 and 96 L/m^2/hr (25 and 57 gfd), as shown in Figure 4.63. In order to maintain constant conditions during the microbial challenge trials it was decided that filtrate samples should be taken after 5 HRTs to ensure that the T_{99} had been achieved prior to sampling.

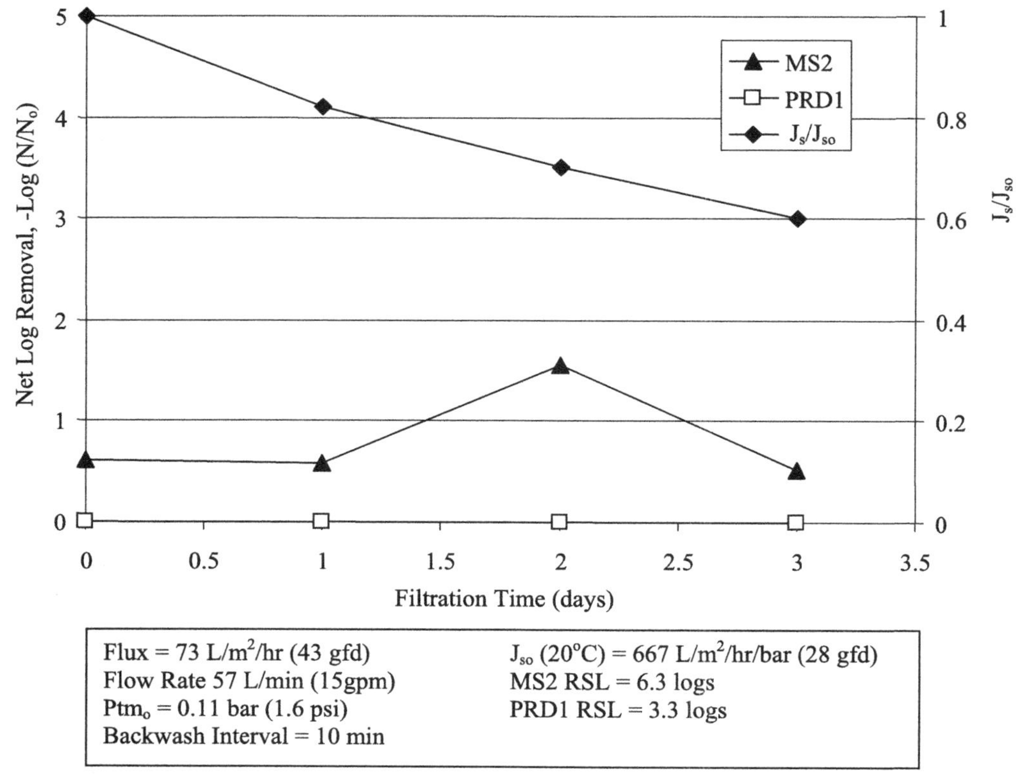

Note:
MS2 bacteriophage removal under clean water conditions at 73 L/m^2/hr = 2.5 logs
PRD1 bacteriophage removal under clean water conditions at 73 L/m^2/hr was at the sensitivity limit of the assay (3.0 logs)

Figure 4.58 Impact of fouling on microbial removal over a 4-day filtration experiment (Trial 1, raw water, submerged PVDF membrane pilot unit, flux = 73 L/m^2/hr, London)

Microbial Removal Efficiency Under Clean Water Conditions

Microbial challenges were performed with the pressurized PES/PVP pilot unit using MS2 and PRD1 bacteriophage seeded into dechlorinated tap water. The filtration flux was initiated at 43 L/m^2/hr (25 gfd), then increased to 70 L/m^2/hr (41 gfd) and finally to 96 L/m^2/hr (57 gfd) in the same filtration cycle. Figure 4.64 shows the average removal observed for MS2 and PRD1 bacteriophage at the three fluxes tested. Each test was performed in triplicate. Error bars indicate the standard deviation of results from the triplicate seeding challenges at each condition.

Evaluation of the Impact of Deposition Layer Formation on Microbial Removal

Raw Water

Microbial challenges were performed at three different fluxes using raw water in the pressurized PES/PVP pilot unit. Six challenges were performed over 4 hours with no backwash. Problems were encountered with maintaining the flux of 96 L/m^2/hr (57 gfd) so experiments at this flux were discontinued after three attempts. Water quality observed at each flux is shown in

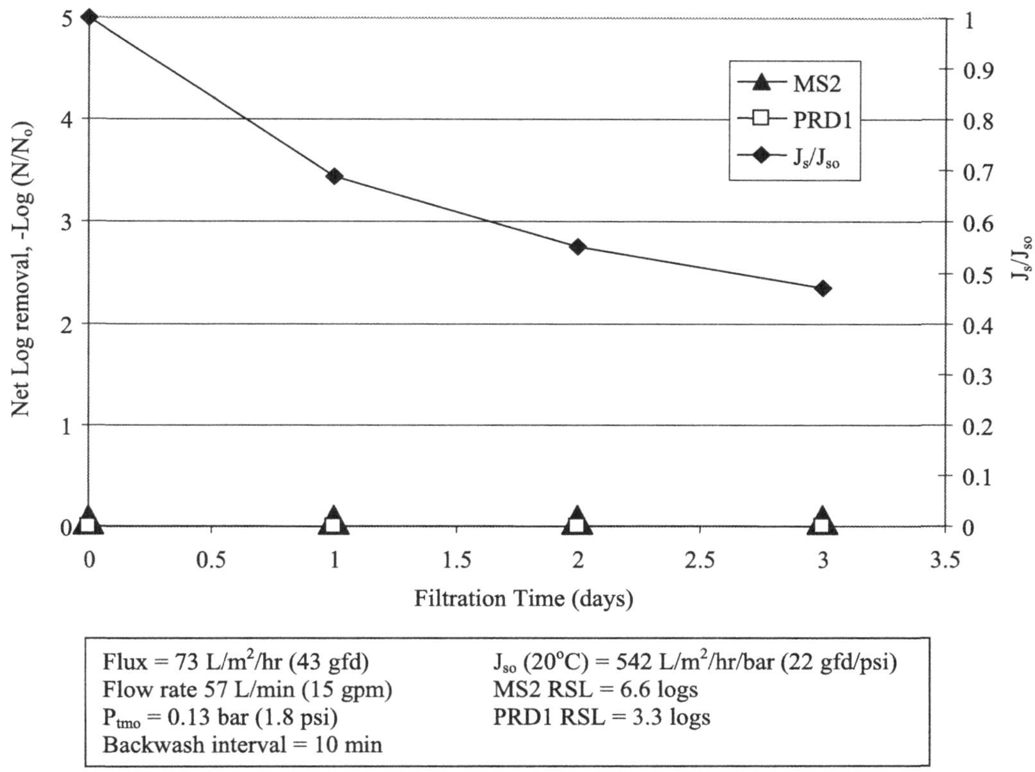

Note:
MS2 removal under clean water conditions at 73 L/m^2/hr = 2.5 logs
PRD1 removal under clean water conditions at 73 L/m^2/hr was at the sensitivity limit of the assay (3.0 logs)

Figure 4.59 Impact of fouling on microbial removal over a 4-day filtration experiment (Trial 2, raw water, submerged PVDF membrane pilot unit, flux = 73 L/m^2/hr, London)

Table 4.27
Filtered water quality during membrane fouling challenges (submerged PVDF membrane pilot unit, London)

Parameter	Unit	Average (min-max)
Temperature	°C	21 (21-22)
Turbidity	NTU	<0.1
pH		7.8 (7.6-7.9)
Hardness	mg/L $CaCO_3$	284 (270-295)
Alkalinity	mg/L $CaCO_3$	208 (199-216)
TOC	mg/L	2.9 (2.7-3.1)
TSS	mg/L	3.4 (<2.0-5.5)
UV_{254}	cm^{-1}	0.034 (0.032-0.037)

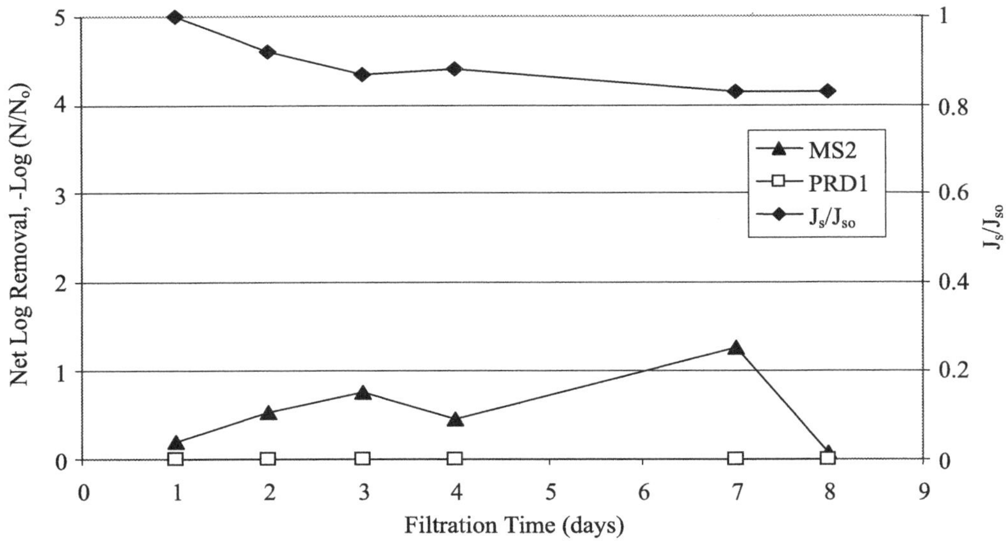

Flux = 73 L/m²/hr (43 gfd)
Flow rate 57 L/min (15 gpm)
P_{tmo} = 0.13 bar (1.9 psi)
Backwash interval = 10 min
J_{so} (20°C) = 542 L/m²/hr/bar (22 gfd/psi)
MS2 RSL = 6.5 logs
PRD1 RSL = 3.4 logs

Note:
MS2 bacteriophage removal under clean water conditions at 73 L/m²/hr = 2.5 logs
PRD1 bacteriophage removal under clean water conditions at 73 L/m²/hr was at the sensitivity limit of the assay (3.0 logs)

Figure 4.60 Impact of fouling on microbial removal over a 4-day filtration experiment (finished water, submerged PVDF membrane pilot unit, flux = 73 L/m²/hr, London)

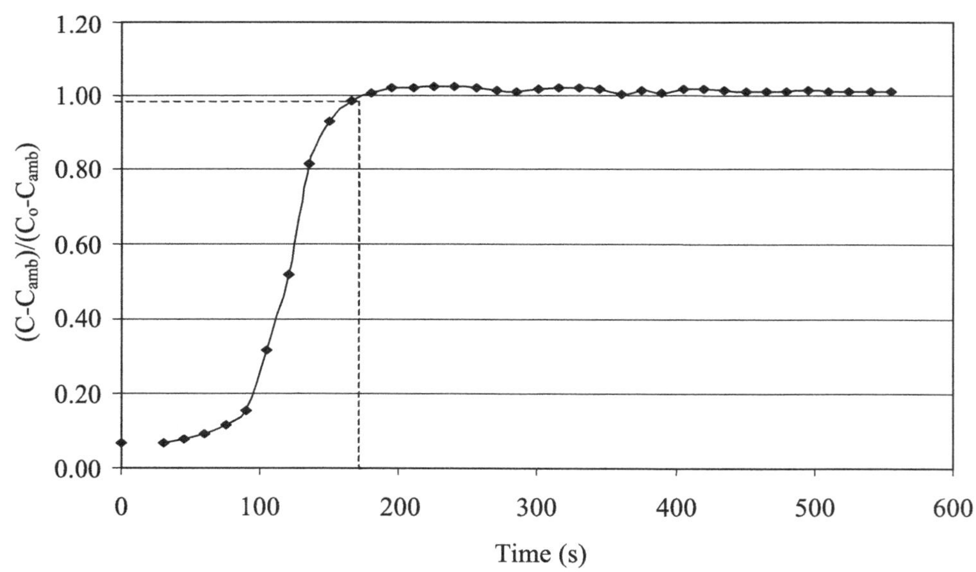

Flow rate = 82 L/min (22 gpm)
Flux = 70 L/m²/hr (41 gfd)

Figure 4.61 Tracer testing results at 70 L/m²/hr (pressurized PES/PVP membrane pilot unit, London)

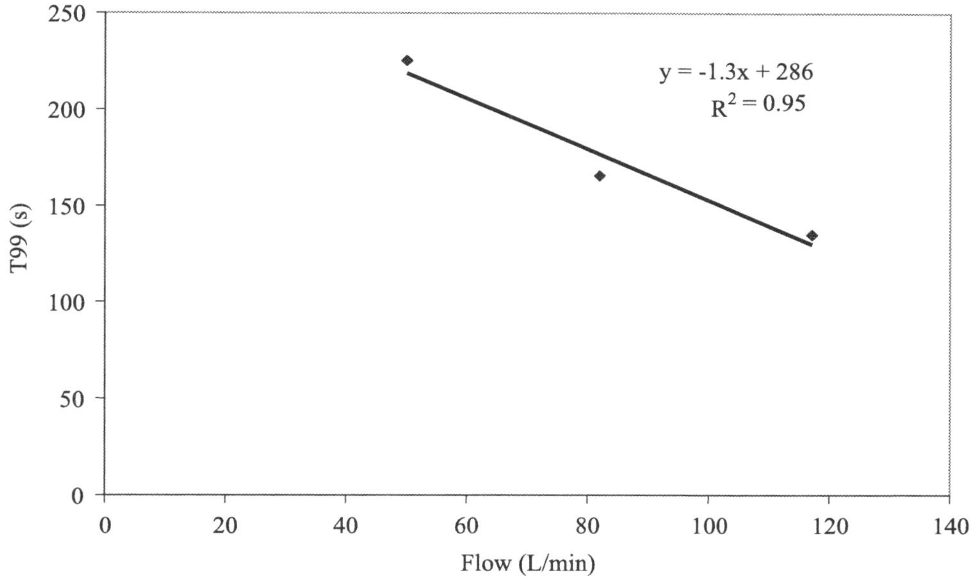

Figure 4.62 T_{99} **as a function of flow rate (pressurized PES/PVP membrane pilot unit, London)**

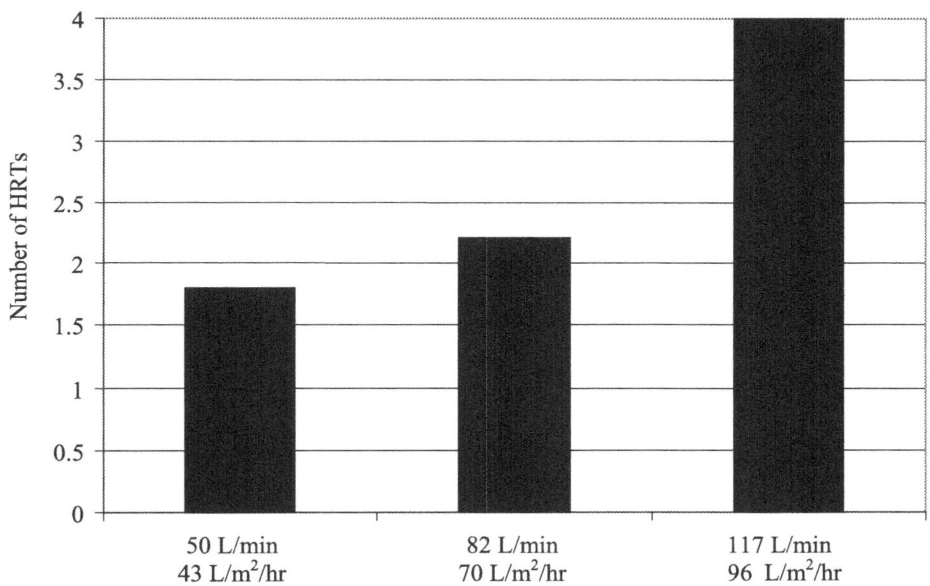

Figure 4.63 **Number of HRTs to reach** T_{99} **(pressurized PES/PVP membrane pilot unit, London)**

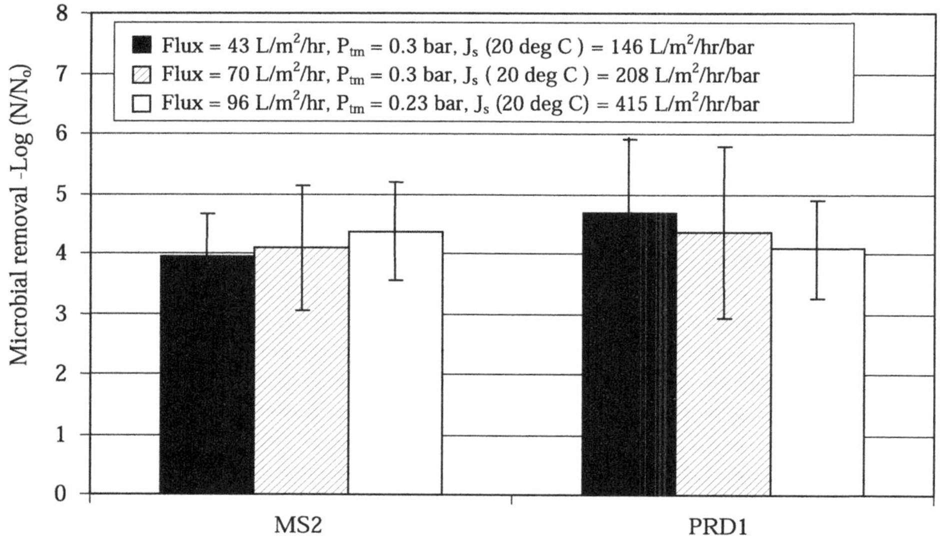

pH = 7.5 | UV_{254} =0.03cm^{-1}
Temperature = 21°C | MS2 RSL = 7.5 logs
Turbidity = 0.18 NTU | PRD1 RSL = 6.6 logs
TOC = 2.2 mg/L

Figure 4.64 Microbial removal under clean water conditions (pressurized PES/PVP membrane pilot unit, London)

Table 4.28. Compared with to removal observed under clean water conditions, no additional removal of MS2 and PRD1 bacteriophage was observed at a flux of 43 L/m^2/hr (25 gfd) as shown in Figure 4.65. Additional removal of MS2 bacteriophage compared to clean water conditions at 70 L/m^2/hr (41 gfd) ranged from 0.0 to 1.9 logs as shown in Table 4.29; no additional removal of PRD1 bacteriophage was observed at that flux.

Filtered Water

Microbial challenges were performed with filtered water in the pressurized PES/PVP pilot unit at three fluxes (43, 71 and 96 L/m^2/hr; 25, 41 and 57 gfd). No additional removal of MS2 and PRD1 bacteriophage compared to clean water conditions was observed over the 4-hour experiment. Water quality recorded during these experiments is shown in Table 4.30. Figure 4.66 illustrates the net microbial removal and specific flux decline observed at a flux of 43 L/m^2/hr (25 gfd). Additional removal was <0.1 log compared to that observed under clean water conditions. Table 4.31 summarizes the net microbial removal and specific flux decline occurring at each flux.

Evaluation of the Impact of Membrane Fouling on Microbial Removal

Raw Water

The pressurized PES/PVP pilot unit was operated for a 5-day period at a constant flux of 70 L/m^2/hr (41 gfd). The length of the test was determined by the ability to maintain the flux through the pressurized PES/PVP pilot unit. The backwash interval was increased to 10 minutes

Table 4.28
Raw water quality during deposition layer formation challenges (pressurized PES/PVP membrane pilot unit, London)

Parameter	Unit	Flow rate 38 L/min Flux 49 $L/m^2/hr$	Flow rate 57 L/min Flux 73 $L/m^2/hr$
Temperature	°C	12	14
Turbidity	NTU	3.0	3.2
pH		8.2	8.3
Hardness	mg/L $CaCO_3$	260	291
Alkalinity	mg/L $CaCO_3$	228	218
TOC	mg/L	3.8	3.8
TSS	mg/L	2.5	5.0
UV_{254}	cm^{-1}	0.07	0.07

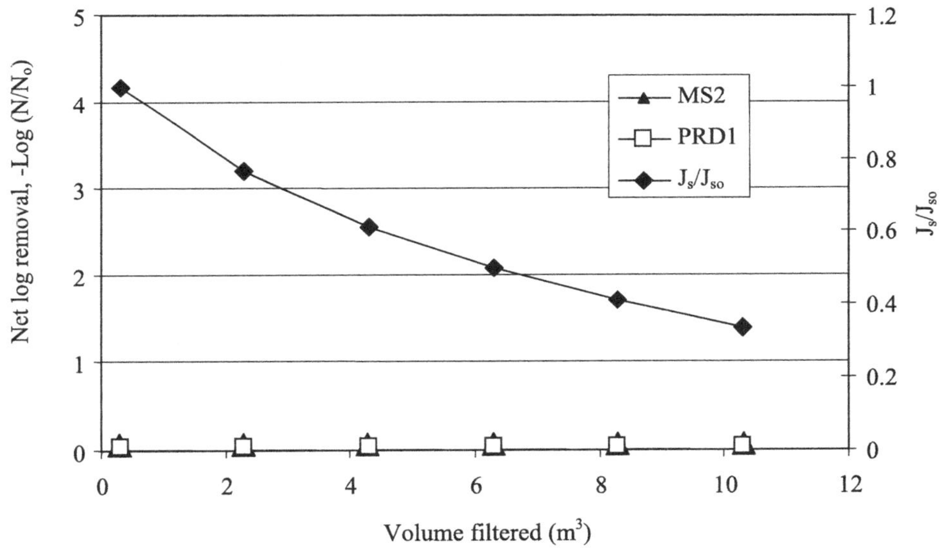

Flux = 43 $L/m^2/hr$ (25 gfd)	J_{so} (20°C) = 259 $L/m^2/hr/bar$ (11 gfd/psi)
Flow rate = 50 L/min (13gpm)	MS2 RSL = 6.3 logs
P_{tmo} = 0.2 bar (2.7 psi)	PRD1 RSL = 7.1 logs

Note:
MS2 bacteriophage removal under clean water conditions at 43 $L/m^2/hr$ = 4.0 logs
PRD1 bacteriophage removal under clean water conditions at 43 $L/m^2/hr$ = 4.7 logs

Figure 4.65 Impact of deposition layer formation on microbial removal (raw water, pressurized PES/PVP membrane pilot unit, flux = 43 $L/m^2/hr$, London)

Table 4.29
Additional microbial removal (compared to clean water conditions) observed during a 4-hour filtration cycle (raw water, pressurized PES/PVP membrane pilot unit, London)

	Flow rate 50 L/min Flux 43 $L/m^2/hr$	Flow rate 82 L/min Flux 71 $L/m^2/hr$
Total volume filtered	12 m^3	20 m^3
Specific flux decline after 4 hours of filtration	67%	77%
Additional removal of MS2 bacteriophage	<0.1 log	<0.1 to 1.9 log
Additional removal of PRD1 bacteriophage	<0.1 log	<0.1 log

Table 4.30
Finished water quality during deposition layer formation challenges (pressurized PES/PVP membrane pilot unit, London)

Parameter	Unit	Flow rate 50 L/min Flux 43 $L/m^2/hr$	Flow rate 82 L/min Flux 70 $L/m^2/hr$	Flow rate 117 L/min Flux 96 $L/m^2/hr$
Temperature	°C	17	15	13
Turbidity	NTU	<0.1	<0.1	<0.1
pH		7.8	7.5	7.9
Hardness	mg/L $CaCO_3$	283	279	292
Alkalinity	mg/L $CaCO_3$	159	140	214
TOC	mg/L	2.1	2.0	2.1
TSS	mg/L	2.4	<2.0	2.0
UV_{254}	cm^{-1}	0.04	0.04	0.02

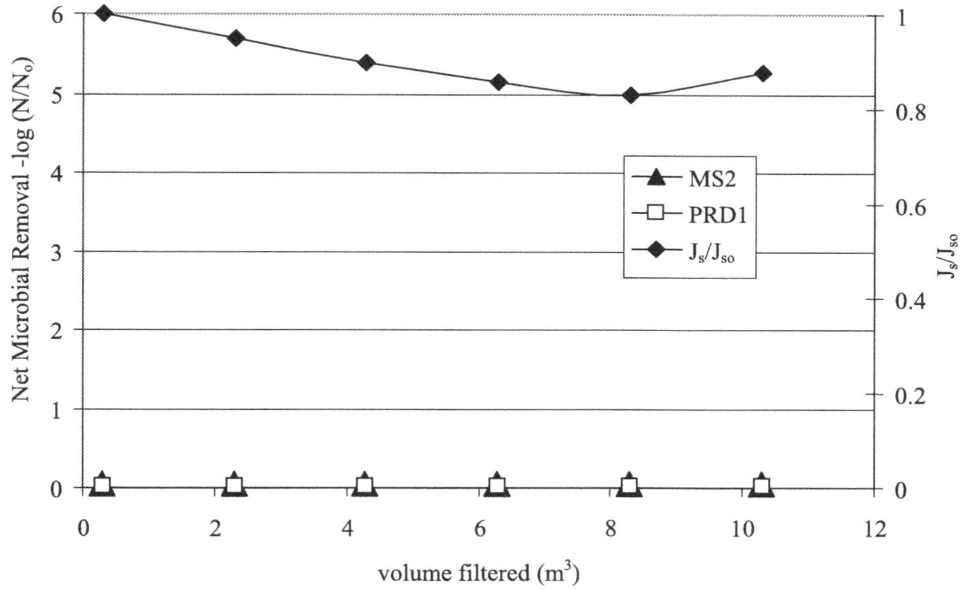

Flux = 43 $L/m^2/hr$ (25 gfd)	J_{so} (20°C) = 353 $L/m^2/hr/bar$ (14 gfd/psi)
Flow rate = 50 L/min (13 gpm)	MS2 RSL = 6.0 logs
P_{tmo} = 0.1 bar (1.4 psi)	PRD1 RSL = 7.0 logs

Note:
MS2 bacteriophage removal under clean water conditions at 43 $L/m^2/hr$ = 4.0 logs
PRD1 bacteriophage removal under clean water conditions at 43 $L/m^2/hr$ = 4.7 logs

Figure 4.66 Impact of deposition layer on microbial removal (filtered water, pressurized PES/PVP membrane pilot unit, flux = 43 $L/m^2/hr$, London)

Table 4.31
Additional microbial removal (compared to clean water conditions) observed during a 4-hour filtration cycle (filtered water, pressurized PES/PVP membrane pilot unit, London)

	Flow rate 50 L/min Flux 43 $L/m^2/hr$	Flow rate 82 L/min Flux 71 $L/m^2/hr$	Flow rate 117 L/min Flux 96 $L/m^2/hr$
Total volume filtered	12 m^3	20 m^3	28 m^3
Specific flux decline after 4 hrs of filtration	12%	16%	29%
Additional removal of MS2 bacteriophage	<0.1 log	0.4 to 1.2 logs	<0.1 log
Additional removal of PRD1 bacteriophage	<0.1 log	1.1 to 3.1 logs	<0.1 log

Table 4.32
Raw water quality during membrane fouling challenges (pressurized PES/PVP membrane pilot unit, London)

Parameter	Unit	Average (min-max)
Temperature	°C	11 (10-12)
Turbidity	NTU	1.3 (0.85-1.7)
pH		8.24 (8.2-8.3)
Hardness	mg/L $CaCO_3$	291 (285-296)
Alkalinity	mg/L $CaCO_3$	221 (218-224)
TOC	mg/L	3.9 (3.8-4.0)
TSS	mg/L	5 (2-8)
UV_{254}	cm^{-1}	0.07 (0.06-0.08)

from 15 minutes and was performed for 1 minute instead of 30 seconds. MS2 and PRD1 bacteriophage were seeded into the pilot plant immediately after a backwash. Raw water quality is summarized for the 5-day test in Table 4.32.

Additional removal of MS2 and PRD1 bacteriophage during the raw water testing compared to clean water conditions was observed, as illustrated in Figure 4.67. Both bacteriophage were removed up to the sensitivity limit of the assay after 5 days of operation.

Filtered Water

Testing was conducted over an 8-day period using filtered water in the pressurized PES/PVP pilot unit at a flux of 70 L/m^2/hr (41 gfd). Feedwater quality occurring during the test is shown in Table 4.33. As shown in Figure 4.68, specific flux declined by 40% over the 8 days of operation. Additional removal of MS2 bacteriophage compared to clean water conditions was observed after 72 hours of operation, with removal of MS2 bacteriophage up to the sensitivity limit of the assay. Little additional removal of PRD1 bacteriophage compared to clean water conditions was observed over the duration of the test.

FULL-SCALE STUDY

Field testing was conducted at the Saratoga Filtration Plant (SFP) operated by the San Jose (California) Water Company. The SFP receives raw water from Saratoga Creek and subjects it to direct membrane filtration through a pressure-driven membrane system followed by chlorine disinfection. The pressurized pilot plant used previously at Atlanta and Scottsdale was relocated to San Jose, installed and operated side-by-side with the full-scale membrane plant. The pilot- and full-scale tests were performed in accordance with the experimental protocol described in Chapter 2, Methods and Materials.

The objective of the testing at the SFP was to quantify the microbial removal of identical polypropylene (PP) membrane modules operated similarly on a full-scale pressure-driven membrane system and the pressurized membrane pilot plant. The membranes had a manufacturer-reported pore size of 0.2 μm. The specific goals of the field testing were:

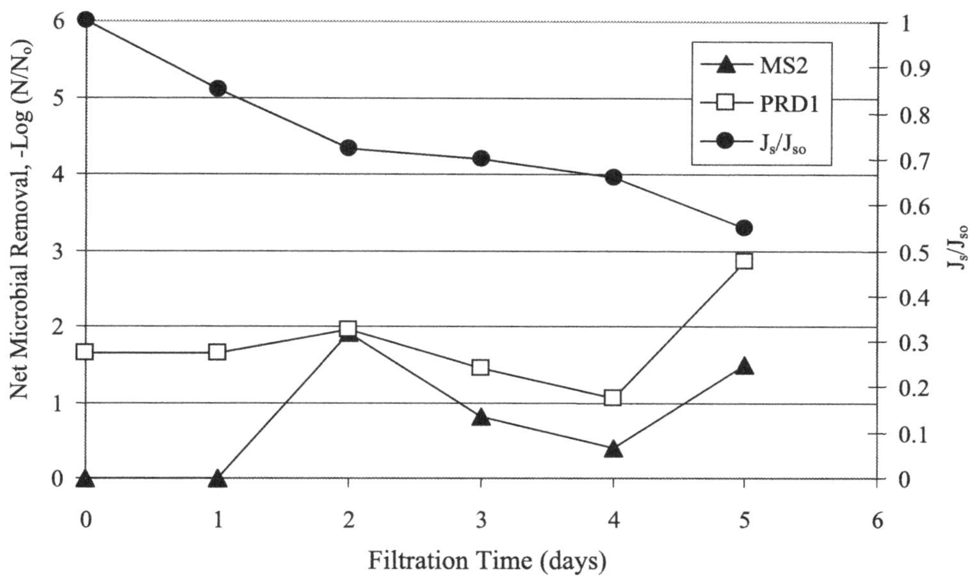

Flux = 70 L/m^2/hr (41 gfd)	J_{so} (20°C) = 369 L/m^2/hr/bar (15 gfd/psi)
Flow rate = 82 L/min (22 gpm)	MS2 RSL = 4.0 to 5.6 logs
P_{tmo} = 0.23 bar (3.3 psi)	PRD1 RSL = 5.0 to 7.2 logs

Note:
MS2 bacteriophage removal under clean water conditions at 70 L/m^2/hr = 4.1 logs
PRD1 bacteriophage removal under clean water conditions at 70 L/m^2/hr = 4.4 logs

Figure 4.67 Impact of fouling on microbial removal over a 5-day filtration experiment (raw water, pressurized PES/PVP membrane pilot unit, flux = 70 L/m^2/hr, London)

Table 4.33
Filtered water quality during membrane fouling challenges (pressurized PES/PVP membrane pilot unit, London)

Parameter	Unit	Average (min-max)
Temperature	°C	11 (10-12)
Turbidity	NTU	<0.1
pH		7.4 (7.1-7.7)
Hardness	mg/L $CaCO_3$	305 (288-329)
Alkalinity	mg/L $CaCO_3$	200 (183-223)
TOC	mg/L	2.1 (1.9-2.2)
TSS	mg/L	3.7 (<2.0-6)
UV_{254}	cm^{-1}	0.02 (0.015-0.03)

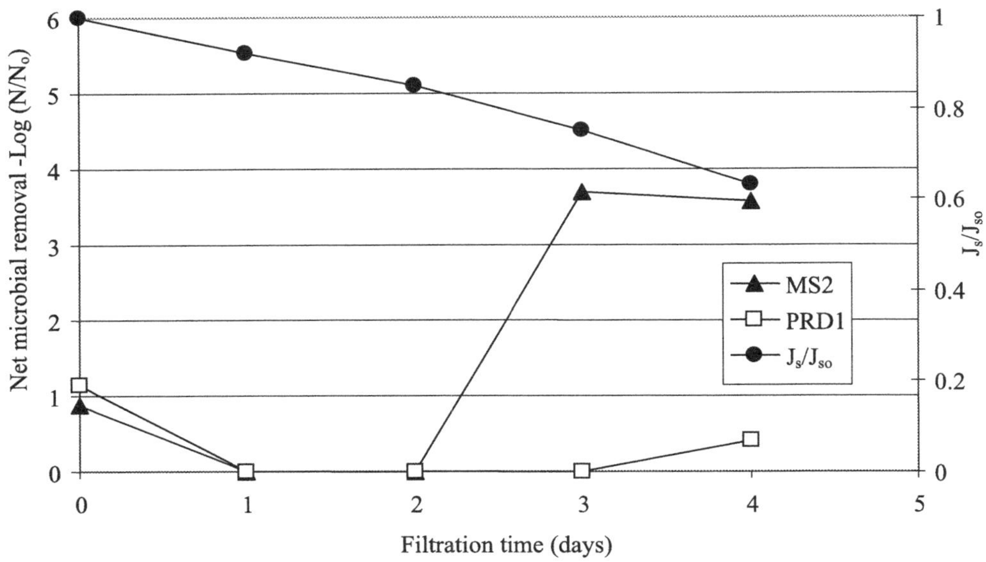

Flux = 70 L/m²/hr (41 gfd)	J_{so} (20°C) = 361 L/m²/hr/bar (15 gfd/psi)
Flow rate = 82 L/min (22 gpm)	MS2 RSL = 7.5 logs
P_{tmo} = 0.23 bar (3.3 psi)	PRD1 RSL = 7.25 logs

Note:
MS2 bacteriophage removal under clean water conditions at 70 L/m²/hr = 4.1 logs
PRD1 bacteriophage removal under clean water conditions at 70 L/m²/hr = 4.4 logs

Figure 4.68 Impact of fouling on microbial removal over an 8-day filtration experiment (finished water, pressurized PES/PVP membrane pilot unit, flux = 70 L/m²/hr, London)

- monitor and compare fouling rates of membrane modules of the same age (brand-new at the beginning of the period of testing) operated on a pilot-scale unit and a full-scale system;
- quantify the microbial removal of these two modules over the period of testing;
- compare the microbial removal of brand-new modules to that of mature modules.

Three modules were used during the field testing: one mature PP module that had been in use on the full-scale plant for five years, and two brand-new PP modules. One brand-new PP module was installed in the pilot plant (pilot-scale new module or PSN) and the two other PP modules were used on the full-scale system (full-scale new module or FSN, and full-scale mature module or FSM).

The pilot- and full-scale units were operated concurrently for nine weeks under the same operating conditions. The modules on the pilot unit were continuously operated at a similar flux to those modules installed on the full-scale plant; and backwashing and chemical cleaning were performed on both units at the same frequency.

Every two weeks, the FSN and FSM modules were removed from the full-scale membrane system and set up on the pressurized membrane pilot unit in order to conduct a microbial challenge. Modules installed in the pilot unit could be isolated from filtration, which allowed challenging the modules individually. After being challenged with microorganisms, the modules were replaced on the full-scale system.

Table 4.34
Raw water quality during full-scale testing (pressurized pilot unit, San Jose)

Parameter	Average (min-max)
Water Temperature (°C)	10 (10-13)
pH	8.0 (7.9-8.2)
Turbidity (NTU)	1.5 (0.8-2.0)
TOC (mg/L)	1.4 (1.2-1.8)
UV_{254} (cm^{-1})	0.03 (0.02-0.04)
Alkalinity (mg/L as $CaCO_3$)	164 (154-174)
Total Hardness (mg/L as $CaCO_3$)	182 (152-203)
Total Ca (mg/L)	50 (41-56)
Total Mg (mg/L)	16 (12-16)

Testing was conducted with raw surface water from Saratoga Creek in Saratoga, CA. Table 4.34 summarizes the raw water quality during the period of testing.

Comparison of Pilot- and Full-Scale Operational Performance

Flux on the pressurized membrane pilot unit was adjusted every day to match the flux of the full-scale membrane system. Typical operating flux ranged from 30 to 45 L/m^2/hr (18 to 27 gfd). Figure 4.69 depicts the evolution of specific flux over the nine-week period for the pilot unit equipped with the PSN module and the full-scale membrane rack where the FSN and FSM modules had been installed.

The specific flux of the PSN module installed on the pilot unit decreased rapidly over the first two weeks. This can be explained by the fact that the PSN and FSN modules were brand-new at the beginning of the field testing. The specific flux of the pilot unit then stabilized and decreased at a slower rate. After the first chemical cleaning, the specific flux variations recorded on the pilot-scale PP membrane unit were similar to the variations monitored on the full-scale PP membrane system. The difference in specific flux between the two systems remained less than 25%. It should be noted that the FSN and FSM modules were two of 90 membrane modules comprising the full-scale membrane rack. Thus, the operating conditions, such as specific flux, of the full-scale membrane rack reflected the operation of all 90 modules, not just the two modules examined in this testing.

Two chemical cleanings (CIP) were performed during the period of the experiment. The first CIP allowed a specific flux recovery of 43% on the full-scale system (from 57 to 81 L/m^2/hr/bar at 20°C; 2.3 gfd/psi to 3.3 gfd/psi at 20°C), whereas the recovery was only 18% on the pilot-scale unit (from 81 to 96 L/m^2/hr/bar at 20°C; 3.3 gfd/psi to 3.9 gfd/psi at 20°C). The chemicals used to perform this first CIP were not optimal since the full-scale membrane rack and the pilot-scale unit were cleaned with solutions that had been used to wash the five other full-scale membrane racks. This resulted in a pH level higher than 5.0 for the cleaning step using citric acid (normal pH approximately 2.0) and a pH level below 10 for the cleaning step using caustic soda

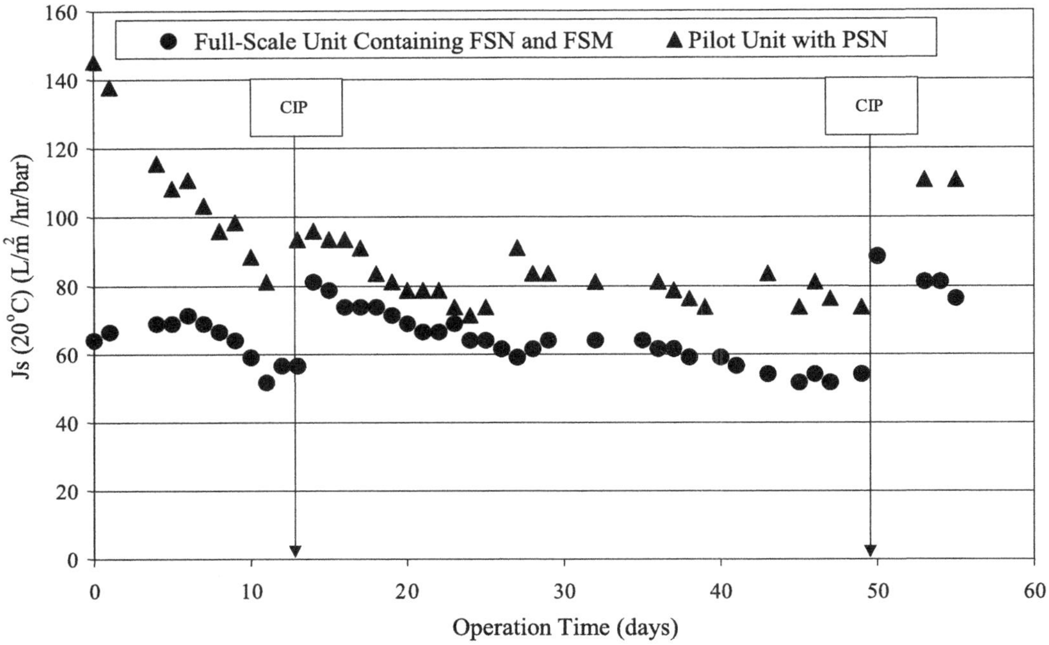

Figure 4.69 Specific flux in pilot- and full-scale pressurized PP membrane units (raw water, San Jose)

(normal pH approximately 12). The second CIP was performed with newer chemical solutions and allowed a specific flux recovery of 63% on the full-scale membrane rack (from 54 to 89 L/m^2/hr/bar at 20°C; 2.2 to 3.6 gfd/psi at 20°C) and 50% on the pilot unit (from 74 to 111 L/m^2/hr/bar at 20°C; 3.0 to 4.5 gfd/psi at 20°C).

Comparison of Pilot- and Full-Scale Microbial Removal Performance

Microbial Removal Efficiency Under Clean Water Conditions

Prior to initiating the nine-week experiment comparing pilot- and full-scale performance, duplicate microbial challenges were conducted using finished (microfiltered but not chlorinated) water from the SFP at a flux of 34 L/m^2/hr (20 gfd) with each of the three PP membrane modules (PSN, FSN and FSM) installed one at a time in the pressurized membrane pilot unit. The goal of these challenges was to quantify the initial removal performance of each PP membrane module under the worst-case conditions from a microbial removal perspective (i.e., clean water conditions with no deposition layer or fouling).

The first microbial challenge was conducted with *P. diminuta* and MS2 and PRD1 bacteriophage seeded into the non-chlorinated microfiltered water from the SFP. Before initiating the microbial feed, a filtrate sample was collected to verify the absence of contamination. *P. diminuta* counts in this sample were higher than 200/100 mL. This was not unexpected since decontamination with chlorine of the pressurized pilot unit could not be performed prior to the test because the PP membranes are not chlorine resistant. Further, the enumeration of *P. diminuta* is subject to interference from other microorganisms due to the use of a non-selective medium in the assay.

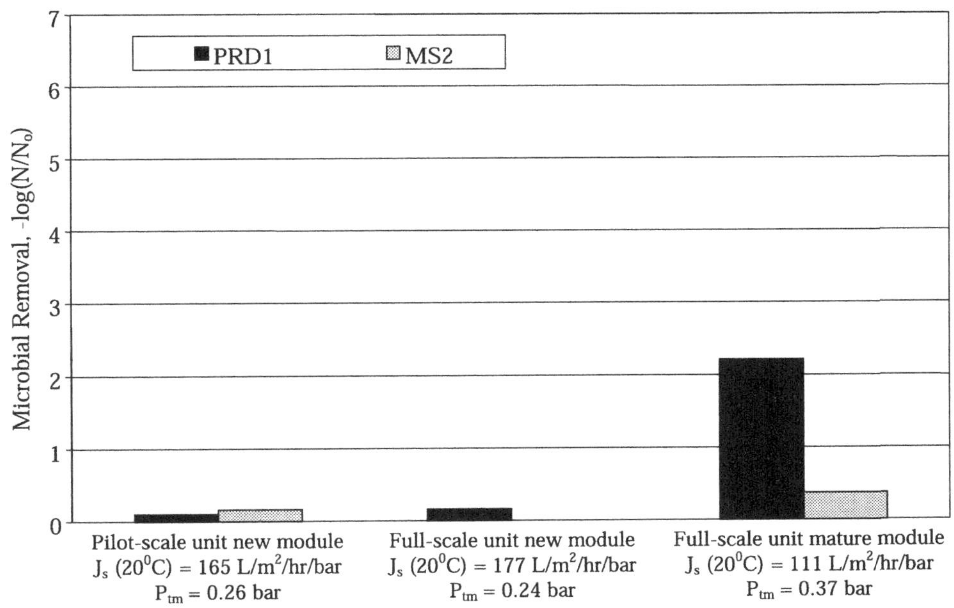

pH = 8.1	UV_{254} = 0.032 cm^{-1}
Temperature = 12°C	MS2 RSL = 7.3 logs
Turbidity = 0.02 NTU	PRD1 RSL = 7.0 logs
TOC = 1.7 mg/L	

Figure 4.70 Microbial removal under clean water conditions for three membrane modules (pressurized PP pilot unit, flux = 34 L/m²/hr, San Jose)

Therefore, no results are presented for removal of *P. diminuta*. Figure 4.70 shows the removal of MS2 and PRD1 bacteriophage observed under clean water conditions for the three PP membrane modules tested.

Microbial removal under clean water conditions was similar for the two new modules (PSN and FSN). Both modules removed a maximum of only 0.2 logs of MS2 and PRD1 bacteriophage. Low microbial removal was expected due to the relatively large membrane pore size (0.2 μm) as compared to the size of the bacteriophage. Removal of both MS2 and PRD1 bacteriophage by the full-scale mature (FSM) module was significantly higher (0.4 and 2.2 logs, respectively) because of its greater maturation and fouling compared to the two new modules. The greater maturation and fouling of the FSM module is supported by the lower specific flux compared to that of the two new modules under similar operating and water quality conditions (110 L/m²/hr/bar for the FSM modules; 165 to 177 L/m²/hr/bar for the two new modules). Removal of MS2 and PRD1 bacteriophage under clean water conditions was used as the baseline to calculate the additional removal induced by deposition layer formation and/or fouling of the PP membranes.

Microbial Removal Efficiency With Raw Water

Four microbial challenges were conducted during the nine weeks of pilot- and full-scale PP membrane testing at San Jose. Every two weeks, the FSN and FSM modules were removed

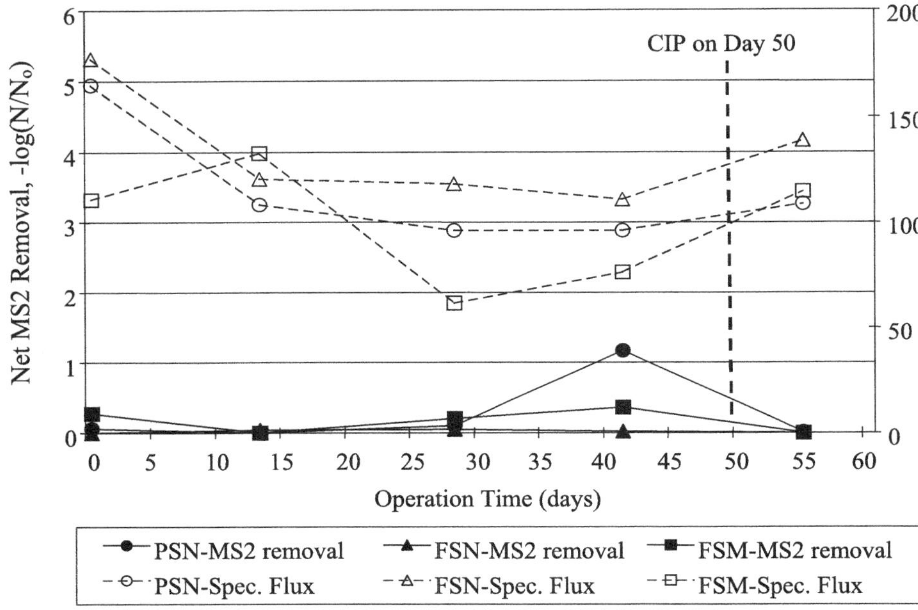

Note:
MS2 bacteriophage removal under clean water conditions at 34 L/m^2/hr = 0.2 log for PSN, 0.0 log for FSN, and 0.4 log for FSM

Figure 4.71 Net MS2 bacteriophage removal for three PP membrane modules (raw water, flux = 34 L/m^2/hr, San Jose)

from the full-scale system and installed one at a time in the pressurized pilot unit in order to conduct a microbial challenge. The challenges were conducted at a flux of 34 L/m^2/hr (20 gfd) using MS2 and PRD1 bacteriophage seeded into the raw water. After being challenged with microorganisms, the FSN and FSM modules were replaced on the full-scale system.

Raw water quality during the testing was presented in Table 4.33. Figures 4.71 and 4.72 depict the additional removal (compared to clean water conditions) of MS2 and PRD1 bacteriophage as a function of the specific flux for the three different PP membrane modules challenged in the pressurized membrane pilot unit.

Very low removal of MS2 bacteriophage was observed from the three modules, ranging from 0.0 to 1.2 logs (Figure 4.71). The additional removal of PRD1 bacteriophage (Figure 4.72) achieved by the two new modules operated on the full-scale (FSN) and pilot-scale (PSN) systems was similar, varying between 0.0 and 1.3 logs. However, the additional removal of PRD1 bacteriophage achieved by the mature module (FSM) ranged from 0.5 to 3.2 logs. The maturity of the module and its level of fouling seemed to substantially impact its microbial removal efficiency. For the same module, a decrease in specific flux resulted in a higher PRD1 bacteriophage removal.

DISCUSSION OF RESULTS FROM ALL PILOT-TESTING LOCATIONS

Pilot testing was performed at three different locations in the US and UK. In Atlanta, testing was performed using a pressurized PVDF membrane pilot plant on raw and rapid-gravity filtered surface water, and a submerged PVDF membrane pilot plant on raw water. In Scottsdale,

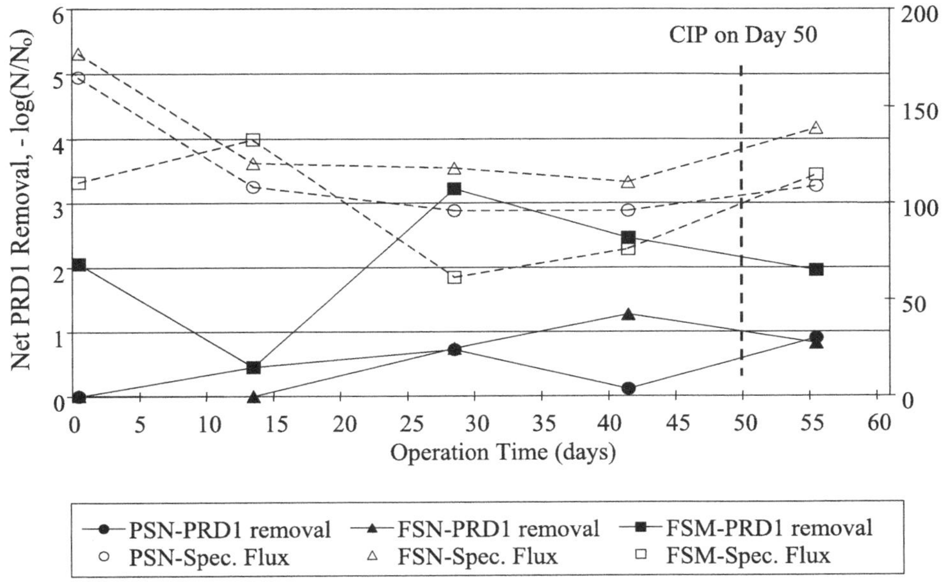

Note:
PRD1 bacteriophage removal under clean water conditions at 34 L/m^2/hr = 0.1 log for PSN, 0.2 log for FSN, and 2.2 logs for FSM

Figure 4.72 Net PRD1 bacteriophage removal for three PP membrane modules (raw water, flux = 34 L/m^2/hr, San Jose)

the pressurized and submerged PVDF membrane pilot plants were tested on secondary effluent from a wastewater treatment facility. In London, a submerged PVDF membrane pilot plant and a pressurized PES/PVP membrane pilot plant was tested using raw and slow-sand filtered surface water. In all locations, pilot testing involved three major tasks that focused on microbial removal under various feedwater and operating conditions. Testing for all tasks was performed at constant flux. The first task focused on the evaluation of microbial removal under clean water conditions, and consisted of microbial challenge tests performed under worst-case conditions with respect to microbial removal (no deposition layer, no fouling, clean feedwater). These tests were performed with dechlorinated tap water. The second task evaluated the impact of deposition layer formation on microbial removal in a series of microbial challenges performed over 4 hours with no backwashing of the membrane to allow the accumulation of a deposition layer. The third task evaluated the impact of membrane fouling on microbial removal by a series of microbial challenges over an extended duration (several days to two weeks), using backwashing but no chemical cleaning to allow fouling constituents to accumulate.

Microbial Removal Under Clean Water Conditions

The first task performed at each pilot-testing location was evaluation of microbial removal under clean water conditions, and consisted of microbial challenge tests performed under worst-case conditions with respect to microbial removal (no deposition layer, no fouling, clean feedwater). These tests were performed with dechlorinated tap water at constant flux. Challenge

Table 4.35
Summary of pilot testing results for microbial removal under clean water conditions

Membrane	Location	Flux (L/m²/hr)	Temperature (°C)	Specific flux (L/m²/hr/bar)	Specific flux at 20°C (L/m²/hr/bar)	PRD1 removal (log_{10})	MS2 removal (log_{10})
PVDF	Atlanta	26	18	153	163	1.6	0.0
pressure		40		160	160	1.4	0.0
		53		212	160	1.2	0.0
PVDF	Atlanta	18	8-13	120	164	0.2	0.4
submerged		27		123	164	0.5	0.3
		37		119	158	0.5	0.2
PVDF	Scottsdale	26	35	260	184	0.4	0.0
pressure		40		267	186	0.2	0.0
		53		265	183	0.3	0.0
PVDF	Scottsdale	18	33	150	108	1.3	1.4
submerged		27		159	108	1.5	1.3
		37		161	106	1.6	1.1
PVDF	London	49	12	700	823	>2.7	3.7
submerged		73		487	599	>3.0	2.7
		86		453	559	>3.0	2.1
PES/PVP	London	43	21	143	146	4.7	4.0
pressure		70		233	208	4.4	4.1
		96		417	415	4.1	4.4

organisms comprised *P. diminuta* bacteria, and MS2 and PRD1 bacteriophage. Results of the clean water testing are summarized in Table 4.35.

The results in Table 4.35 indicate that the pressurized PVDF membrane pilot unit did not provide any removal of MS2 bacteriophage under clean water conditions at either Atlanta or Scottsdale. The manufacturer-reported pore size of this membrane is 0.1 microns, and the size of MS2 bacteriophage is 0.024 microns. However, this membrane did provide removal of PRD1 bacteriophage (0.070 microns), which is also smaller than the manufacturer-reported pore size of the membrane.

Further, PRD1 bacteriophage removal by the pressurized PVDF membrane pilot unit at Atlanta was higher than at Scottsdale for clean water conditions. The clean water testing was performed as the initial pilot-testing task at each location, and Atlanta pilot testing was performed approximately one year earlier than the Scottsdale testing using the same membrane. It may be expected that due to a year of use in Atlanta, the membrane would have been irreversibly fouled to some extent, which would lead to higher microbial removal in Scottsdale than Atlanta under clean water conditions. Due to the substantial feedwater temperature difference between Atlanta and Scottsdale (18°C and 35°C, respectively), specific flux must be normalized for temperature in order to compare results from the two locations. The specific flux was higher at Scottsdale (183 to 186 L/m^2/hr/bar at 20°; 8 gfd/psi) than at Atlanta (160 to 163 L/m^2/hr/bar at 20°; 7 gfd/psi). These results indicate that the membrane had not been subject to irreversible fouling. Further data would be necessary to fully understand the observed differences in microbial removal results.

A different trend was observed in clean water results for the submerged PVDF membrane pilot unit in Atlanta and Scottsdale. The specific flux for this membrane suggested that irreversible fouling had occurred in the approximately 6 months between clean water testing at Atlanta and Scottsdale. The specific flux of the submerged PVDF membrane pilot unit was lower in Scottsdale (106 to 108 L/m^2/hr/bar at 20°; 4 gfd/psi) than at Atlanta (158 to 164 L/m^2/hr/bar at 20°; 7 gfd/psi). Removal of PRD1 bacteriophage at Scottsdale (1.3 to 1.6 logs) compared to Atlanta (0.2 to 0.5 logs). These trends are expected since the submerged PVDF membrane was new at the time pilot testing in Atlanta was initiated. After six months of use with raw and filtered feedwater in Atlanta, membrane maturation and irreversible fouling had occurred to the extent that specific flux was lower in Scottsdale than Atlanta.

Table 4.35 also indicates that although the material and manufacturer-reported pore size of the pressurized and submerged PVDF membranes used at the two pilot-testing locations were identical, their microbial removal performance was different. The submerged PVDF membrane pilot unit removed MS2 bacteriophage at both locations and the pressurized PVDF membrane pilot unit did not remove MS2 bacteriophage at either location under clean water conditions. It should be noted that the differences in removal, regardless of location, membrane system type, or bacteriophage type, were less than 1.6 logs. While these differences may be statistically valid, they may not be discernible from a regulatory perspective.

The clean water microbial removal results shown in Table 4.35 also demonstrate the difference between microbial removal of the two membranes used at London (PVDF submerged and PES/PVP pressurized) compared to the pressurized and submerged PVDF membranes used in the US. The manufacturer-reported MWCOs of the membranes used at London (150 kD or 0.01 microns for the pressurized PES/PVP membrane, and 0.035 microns for the submerged PVDF membrane) were substantially smaller than those of the PVDF membranes used in the US (0.1 microns). As expected based on manufacturer-reported MWCO or pore size, both MS2 and PRD1 bacteriophage were very effectively removed by the two membranes used at London. Removal of PRD1 bacteriophage (0.07 microns) by the pressurized PES/PVP membrane pilot unit under clean water conditions exceeded 4.0 logs, and PRD1 bacteriophage removal by the PVDF submerged membrane pilot unit exceeded the sensitivity limit of the assay (2.7 to 3.0 logs). Although MS2 bacteriophage (0.024 microns) is smaller than the manufacturer-reported pore size of the submerged PVDF membrane (0.035 microns), the membrane removed 2.1 to 3.7 logs of MS2 bacteriophage under clean water conditions. For the pressurized PES/PVP membrane (manufacturer-reported MWCO of 150 kD or 0.01 microns), MS2 removal was even more effective (4.0 to 4.4 logs).

Impact of Deposition Layer Formation on Microbial Removal

The second task performed at each pilot-testing location was an evaluation of the impact of deposition layer formation on microbial removal. At each location, a series of microbial challenges was performed over 4 hours. The membranes were not backwashed during this time in order to allow the accumulation of a deposition layer. MS2 and PRD1 bacteriophage were seeded into the feedwater, which was either raw water or filtered water at Atlanta and London, or secondary wastewater effluent at Scottsdale. The net microbial removal was calculated, defined as the additional removal of MS2 and PRD1 bacteriophage in the deposition layer formation challenges compared to that observed under clean water conditions. The net microbial removal was that attributable to deposition layer formation during the 4-hour experiment. Results of the deposition layer formation challenges from all pilot-plant locations and feedwaters are summarized in Table 4.36.

For Atlanta and Scottsdale, as expected net PRD1 bacteriophage removal was consistently higher than net MS2 bacteriophage removal. Results from the UK pilot testing indicate that there was very little removal of either MS2 or PRD1 bacteriophage as a result of deposition layer formation. This observation is consistent with the very high levels of bacteriophage removal with the pressurized PES/PVP membrane and the submerged PVDF membrane occurring in clean water testing at London. Because bacteriophage removal was so high under clean water conditions, additional removal due to deposition layer formation was negligible under most conditions tested with both membranes at London.

In Atlanta, there was little consistent effect of deposition layer formation on MS2 bacteriophage removal in either raw or filtered water, regardless of membrane type (pressurized or submerged PVDF membranes). Example results for the pressurized PVDF membrane using raw water are shown in Figure 4.73. The figure indicates that specific flux decline was generally less than 20% over the 4-hour experiment, suggesting that the particulate and organic content of the raw feedwater was insufficient to cause any significant deposition layer formation. With low deposition on the membrane, little additional microbial removal was observed compared to clean water conditions.

The relationship between specific flux decline and PRD1 removal was only slightly more pronounced for the submerged PVDF membrane operated with raw water in Atlanta (flux decline ranging from 28 to 38% and net PRD1 bacteriophage removal ranging from 0.7 to 1.8 logs over the 4-hour experiment). This relationship is shown in Figure 4.74.

For Atlanta filtered water, the flux decline was almost nonexistent (less than 5% over 4 hours), leading to no additional MS2 bacteriophage removal and minimal additional PRD1 bacteriophage removal (<0.4 logs and up to 1.3 logs, respectively).

With secondary wastewater effluent as the feedwater in Scottsdale, specific flux declines were greater than those observed in Atlanta (29 to 50% for the pressurized PVDF membrane and 51 to 55% for the submerged PVDF membrane). However, net removal of both MS2 and PRD1 bacteriophage was only slightly higher in Scottsdale than Atlanta. Example results are shown in Figure 4.75 for the submerged PVDF membrane pilot unit in Scottsdale. Flux declines of up to 55% led to additional removal of PRD1 bacteriophage ranging from 0.7 to 2.0 logs compared to clean water conditions. Net removal of MS2 bacteriophage was not as great as that of MS2, ranging from 0.4 to 1.3 logs for the submerged PVDF membrane. For the pressurized PVDF membrane, deposition layer formation produced specific flux declines up to 50%, and led to the highest and most consistent net removal of PRD1 bacteriophage, from 0.2 to 3.1 logs over the 4-hour experiment. Net removal of MS2 bacteriophage by this membrane ranged from 0.0 to 1.6 logs.

Table 4.36
Summary of pilot testing results for the impact of deposition layer formation on microbial removal

Membrane	Location	Flux (L/m^2/hr)	Specific flux decline (%) over 4 hours	Net PRD1 removal range ($\log_{10}$)	Net MS2 removal range ($\log_{10}$)
Pressurized PVDF	Atlanta	26	13	0.6 to 1.0	0.1 to 0.7
raw water		40	19	0.8 to 1.3	0.0 to 1.3
		53	14	0.6 to 1.3	0.2 to 0.8
Pressurized PVDF	Atlanta	26	1	<0.5	<0.4
filtered water		40	3	0.0 to 1.3	<0.1
		53	3	0.0 to 0.7	<0.1
Submerged PVDF	Atlanta	11	28	0.7 to 1.5	0.0
raw water		27	38	0.9 to 1.8	0.0 to 1.1
		37	33	1.4 to 1.8	0.1 to 0.6
Pressurized PVDF	Scottsdale	26	29	0.2 to 2.4	0.0 to 0.8
secondary effluent		40	50	0.7 to 2.6	0.2 to 0.7
		53	50	1.1 to 3.1	0.5 to 1.6
Submerged PVDF	Scottsdale	18	55	1.1 to 2.0	0.4 to 1.3
secondary effluent		27	53	0.7 to 1.5	0.6 to 0.9
		37	51	1.3 to 1.9	0.7 to 1.0
Submerged PVDF	London	48	1	<0.1	0.0 to 1.0
raw water		73	25	<0.1	<0.1
		86	26	<0.1	<0.1
Submerged PVDF	London	49	2	<0.1	<0.1
filtered water		73	3	<0.1	0.0 to 0.1
		86	8	<0.1	0.5 to 1.9
Pressurized PES/PVP	London	43	67	<0.1	<0.1
raw water		71	77	<0.1	<0.1 to 1.9
Pressurized PES/PVP	London	43	12	<0.1	<0.1
filtered water		71	16	1.1 to 3.1	0.4 to 1.2
		96		<0.1	<0.1

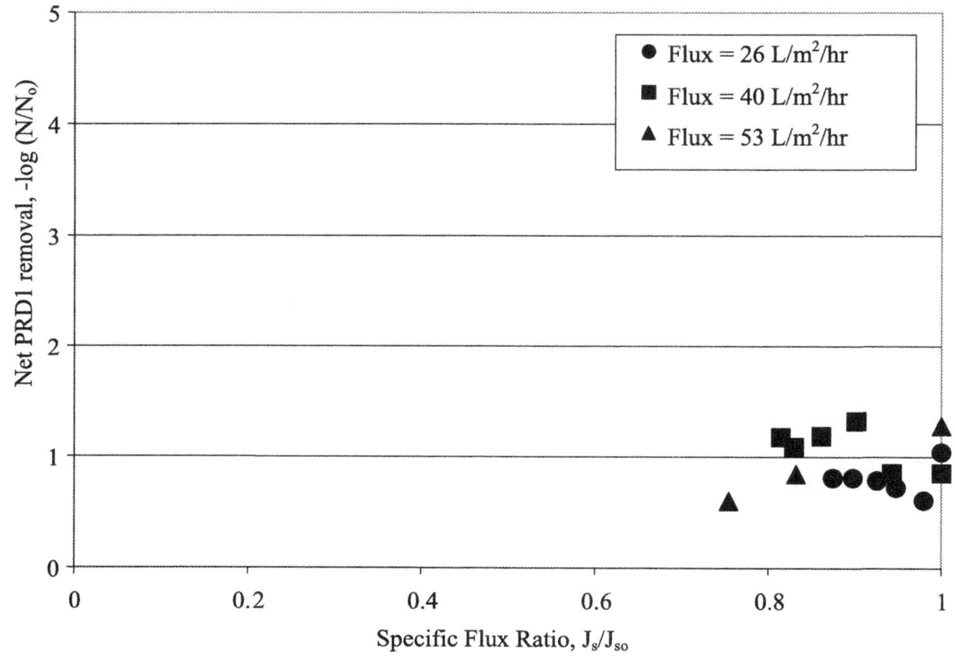

Figure 4.73 PRD1 bacteriophage removal as a function of specific flux decline during deposition layer formation challenges (raw water, pressurized PVDF membrane pilot unit, Atlanta)

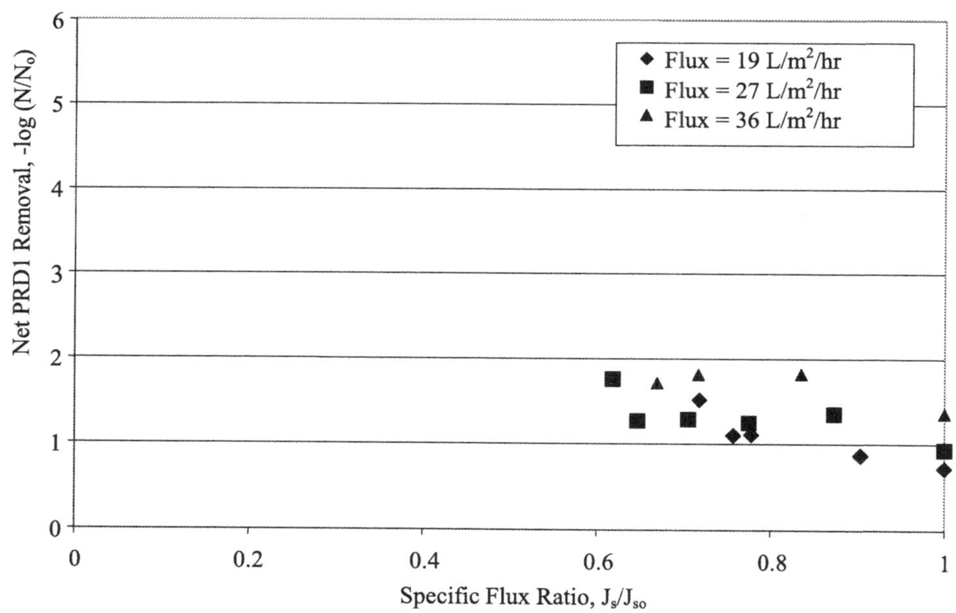

Figure 4.74 PRD1 bacteriophage removal as a function of specific flux decline during deposition layer formation challenges (raw water, submerged PVDF membrane pilot unit, Atlanta)

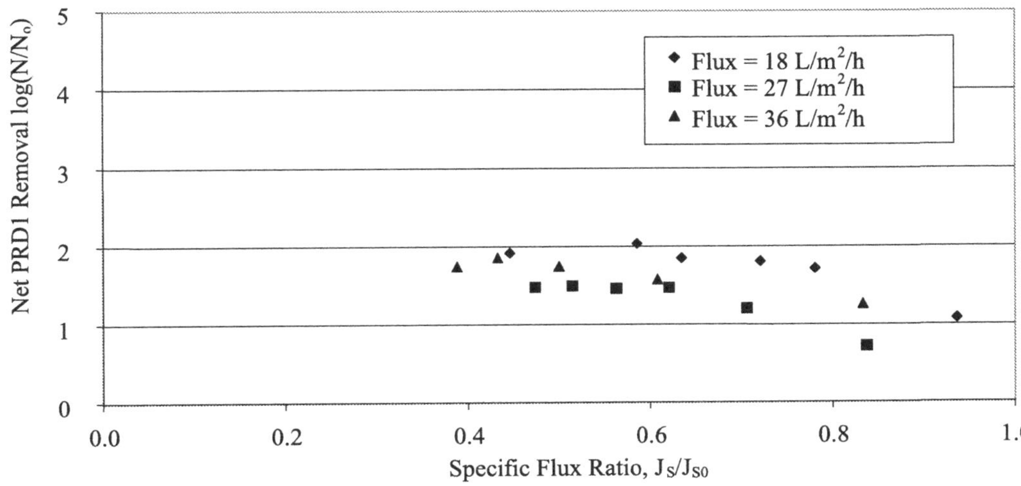

Figure 4.75 PRD1 bacteriophage removal as a function of specific flux decline during deposition layer formation challenges (secondary effluent, submerged PVDF membrane pilot unit, Scottsdale)

In summary, it appears that a deposition layer on the membrane may contribute, at most, an additional 2 logs removal of virus. It should be noted, however, that most low pressure membranes do not operate under the extreme conditions employed here (i.e., four hours without backwashing). Thus, most full-scale plants will experience less than two logs additional removal under normal operating conditions.

Impact of Membrane Fouling on Microbial Removal

The third task performed at each pilot plant location evaluated the impact of membrane fouling on microbial removal. For the purposes of this study, membrane fouling was considered as an observed decline in specific flux. Deposition layer did not contribute to the fouling as challenge experiments were conducted immediately after backwashing the membrane. A series of microbial challenges was conducted over an extended period of membrane operation (several days to two weeks), using backwashing but no chemical cleaning to allow fouling constituents to accumulate. Feedwaters included raw and filtered water at Atlanta and London, and secondary wastewater effluent at Scottsdale. Experiments were terminated after two weeks or after the maximum trans-membrane pressure had been reached. Results of the membrane fouling tests at all three pilot locations are summarized in Table 4.37.

The data in Table 4.37 indicate that membrane fouling had a much greater influence on microbial removal than did deposition layer formation under the conditions tested. The trend is illustrated in Figure 4.76. Flux declines ranged from 30 to 90% for all but one of the fouling experiments. In all tests at Atlanta, membrane fouling led to increased removal of PRD1 bacteriophage, regardless of feedwater (raw or filtered) and membrane type (pressurized or submerged PVDF membranes). At the end of the experiment, maximum additional PRD1 bacteriophage

Table 4.37
Summary of pilot testing results for the impact of membrane fouling on microbial removal

Membrane, feedwater type and duration of experiment	Location	Flux ($L/m^2/hr$)	Specific flux decline (%) during experiment	Net PRD1 removal (log_{10})	Net MS2 removal (log_{10})
Pressurized PVDF/raw Duration: 2 weeks	Atlanta	40	58	2.7 to 4.9	0.2 to 5.0
Pressurized PVDF/filtered Duration: 2 weeks	Atlanta	53	31	0.0 to 1.6	0.0 to 0.8
Submerged PVDF/raw Trial 1 Duration: 30 hours	Atlanta	27	60	1.9 to 3.2	0.2 to 0.9
Submerged PVDF/raw Trial 2 Duration: 11 days	Atlanta	18	58	0.9 to 4.1	0.2 to 0.9
Pressurized PVDF/secondary effluent Duration: 2 weeks	Scottsdale	40	90	0.0 to 5.4	0.3 to 4.4
Submerged PVDF/secondary effluent Duration: 2 weeks	Scottsdale	18	30	1.3 to 3.1	0.0 to 0.8
Submerged PVDF/raw Trial 1 Duration: 4 days	London	73	40	0.0	0.5 to 1.6
Submerged PVDF/raw Trial 2 Duration: 4 days	London	73	50	0.0	0.0
Submerged PVDF/filtered Duration: 8 days	London	73	17	0.0	0.0 to 1.3
Pressurized PES/PVP/raw Duration: 5 days	London	70	45	1.1 to 2.9	0.0 to 2.0
Pressurized PES/PVP/filtered Duration: 5 days	London	70	37	0.0 to 1.1	0.0 to 3.7

removal compared to clean water conditions ranged from 1.6 to 4.9 logs at Atlanta. Additional removal of MS2 bacteriophage was not as consistent as that of PRD1, ranging from 0.0 to 0.9 logs for raw and filtered water for the submerged PVDF membrane and the pressurized PVDF membrane using filtered water, but reached 5.0 logs for the pressurized PVDF membrane using raw water. In Scottsdale pilot testing using secondary effluent as feedwater, maximum flux decline over the 2-week experiments ranged from 30% for the submerged PVDF membrane, producing a maximum net PRD1 bacteriophage removal of 4.1 logs, to 90% for the pressurized PVDF membrane, which produced a maximum net PRD1 removal of 5.4 logs and a maximum net MS2 bacteriophage removal of 4.4 logs.

The pressurized PVDF membrane was the only one of the two membranes tested for which fouling produced a discernible and consistent trend in MS2 bacteriophage removal, using raw water at Atlanta (up to 5.0 logs at a maximum specific flux decline of 58%) and using secondary effluent at Scottsdale (up to 4.4 logs at a maximum specific flux decline of 90%). Using filtered water, this membrane had a maximum specific flux decline of only 31%, which led to a maximum PRD1 bacteriophage removal of 1.6 logs and a maximum MS2 bacteriophage removal of less than 1.0 log. This membrane appears to have been irreversibly fouled to some extent since after termination the specific flux recovered from a 90% decline to a 60% decline upon chemical cleaning after the two-week test.

Microbial removal with membrane fouling was less consistent in testing at London. Maximum flux decline was 40 to 50% for the submerged PVDF membrane and 45% for the pressurized PES/PVP membrane using raw water. Maximum flux decline using filtered water was only 17% for the submerged PVDF membrane and 37% for the pressurized PES/PVP membrane. The additional removal of PRD1 did not exhibit a consistent trend for either of the membranes or feedwater types in London testing. For MS2, additional removal reached 3.7 logs for the pressurized PES/PVP membrane, but the trend of increasing microbial removal with increasing flux decline was inconsistent.

A comparison of the impact on microbial removal by deposition layer formation and membrane fouling is shown in Figure 4.77 for the pressurized PVDF membrane using raw water at Atlanta. The figure indicates that membrane fouling had a much more substantial and consistent impact on specific flux decline, and hence microbial removal, than did deposition layer formation for the conditions tested.

Even with poorer feedwater quality at Scottsdale (secondary effluent) compared to raw water at Atlanta), membrane fouling had less of an impact on microbial removal. For the secondary effluent feedwater, pilot testing data suggest that membrane fouling had a only slightly greater impact on microbial removal for a given flux decline. These results are illustrated in Figure 4.77.

COMPARISON OF BENCH- AND PILOT-SCALE RESULTS

One of the key objectives of pilot-scale testing in this project was to verify results obtained at bench scale. Bench-scale results for the following issues were tested at the pilot scale:

- Scale-up of microbial removal from bench to pilot
- Impact of manufacturer-reported MWCO or pore size on microbial removal (membrane characteristics)
- Impact of deposition layer formation on microbial removal (water quality)

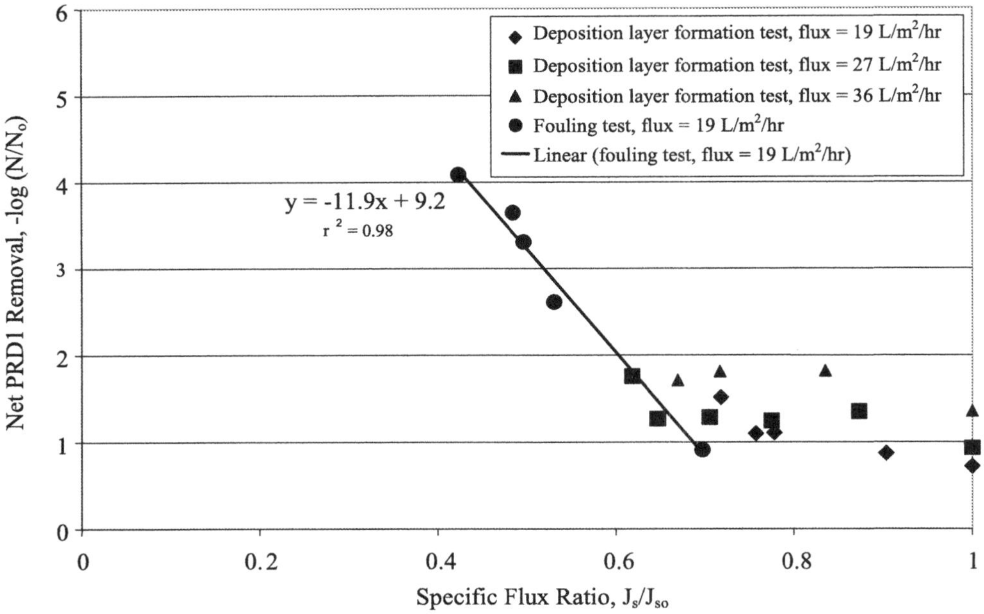

Figure 4.76 PRD1 bacteriophage removal as a function of specific flux decline during deposition layer formation and membrane fouling challenges (raw water, pressurized PVDF membrane pilot unit, Atlanta)

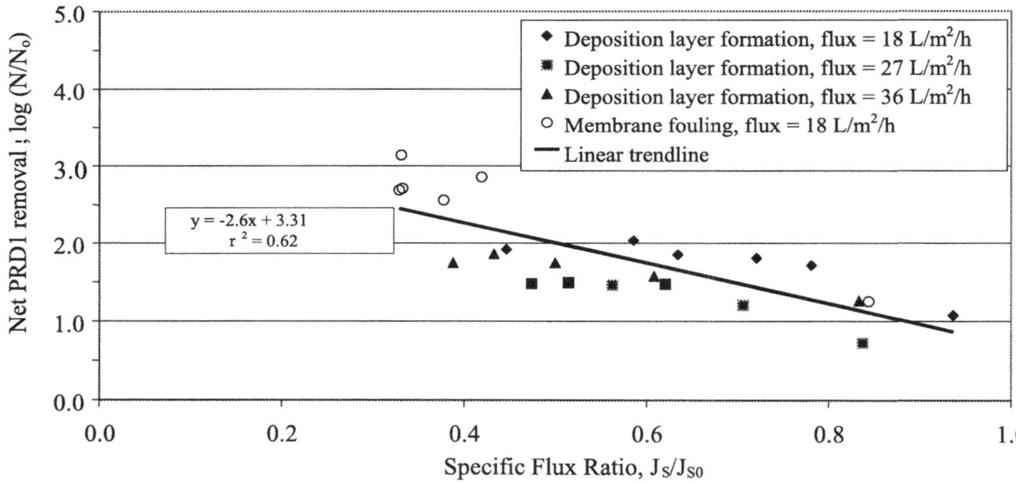

Figure 4.77 PRD1 bacteriophage removal as a function of specific flux decline during deposition layer formation and membrane fouling challenges (secondary effluent, submerged PVDF membrane pilot unit, Scottsdale)

Scale-Up of Microbial Removal From Bench to Pilot

Two of the membranes used in bench-scale testing at JHU were also used in pilot testing, allowing an assessment of the scale-up of microbial removal from bench- to pilot-scale membranes. The membrane designated "PVDF 0.1 μm" had the same membrane type and manufacturer as the "pressurized PVDF membrane pilot unit" employed in pilot testing at Atlanta and Scottsdale. Further, the membrane designated "PVDF 0.035 μm" in bench-scale testing had the same membrane type and manufacturer as the "submerged PVDF membrane pilot unit" employed in pilot testing at London.

A comparison of the PVDF 0.1 μm bench-scale performance and the pressurized PVDF membrane pilot unit was shown in Figures 4.31 and 4.32 for Atlanta using dechlorinated tap water. The microbial removal results from bench and pilot scale are comparable. No removal of MS2 bacteriophage was observed at either bench or pilot scale. Removal of PRD1 bacteriophage by this membrane was 1.2 to 1.6 logs at bench scale and 0.5 to 0.7 log at pilot scale. Specific flux was slightly higher at bench scale than at pilot scale (40 to 79 L/m^2/hr/bar at bench scale; 26 to 53 L/m^2/hr/bar at pilot scale). The difference in specific flux is most likely due to the use of new membranes in bench-scale testing and the fact that these membranes had been subjected to only laboratory water as feedwater until this single set of tests, instead of the natural waters used at pilot scale that contain NOM and other fouling constituents at pilot scale. PRD1 bacteriophage removal was slightly higher in the pilot-scale testing due to the greater maturity and irreversible fouling of the membrane used in pilot-scale testing.

The microbial removal performance of the PVDF 0.035 μm membrane at bench scale can be compared to that of the submerged PVDF membrane pilot unit used in pilot-scale testing at London. These results were presented in Figures 4.1 and 4.56; however, the bench-scale results were obtained using laboratory water and the pilot-scale results were obtained using dechlorinated tap water. However, the MS2 removal results are comparable from both sets of testing. MS2 removal at bench scale was 3.8 logs. MS2 removal at pilot scale ranged from 3.7 logs to 2.3 logs.

Impact of Manufacturer-Reported MWCO or Pore Size on Microbial Removal

The impact of manufacturer-reported MWCO or pore size was examined at bench scale in Figure 4.1 for MS2 bacteriophage removal by a variety of membranes. MS2 bacteriophage is approximately 0.024 μm in size. Based on size exclusion alone, it would be expected that membranes with pore sizes greater than 0.024 μm would allow MS2 to pass into the filtrate, and membranes with pore sizes smaller than 0.024 μm would retain MS2. Figure 4.1 illustrated that a membrane with a manufacturer-reported pore size of 0.035 μm provided effective removal of MS2 bacteriophage (3.8 logs) which is the same level of removal provided by the "tightest" membrane tested (10 kD). However, membranes with manufacturer-reported pore sizes of 300 and 100 kD removed MS2 bacteriophage up to the sensitivity limit of the assay, which was almost 2 logs greater than the removal provided by the 10-kD membrane. Similar trends were illustrated in Figure 4.4, in which removal of MS2 and PRD1 bacteriophage, as well as poliovirus, hepatitis A virus and feline calicivirus, by an mPS membrane with a manufacturer-reported MWCO of 100 kD was higher by approximately 1.0 log than that of an mPS membrane with a manufacturer-reported MWCO of 10 kD. Figure 4.16 provides further evidence of the trend. Removal of MS2 and PRD1 bacteriophage and poliovirus by 100- and 300-kD mPS membranes was as high or higher than that provided by a 10-kD membrane (all MWCOs as reported by the manufacturer).

At pilot scale, the fact that manufacturer-reported pore size or MWCO was not a reliable predictor of microbial removal performance is illustrated. The PVDF pressurized and submerged membrane pilot units had a manufacturer-reported pore size of 0.1 μm. This membrane provided 0.0 to 1.4 logs of MS2 bacteriophage removal, depending on the test conditions. This membrane provided 0.2 to 1.7 logs of PRD1 removal under the same conditions. At bench scale, the same membrane (same membrane material, manufacturer, and manufacturer-reported pore size) provided 0.3 logs of MS2 removal. Thus, a range of microbial removal for the same membrane was observed at pilot scale, thereby verifying the bench-scale results.

Effects of Deposition Layer on Microbial Removal

The effects of a deposition layer on microbial removal were illustrated at bench scale in Figure 4.15. In this figure, increasing the bentonite loading (i.e., increasing the amount of material deposited on the membrane) caused increasing removal of poliovirus as well as MS2 and PRD1 bacteriophage. The increase was clearly discernible from the "no bentonite deposition" (clean water) condition, and was consistent with increasing bentonite load for all three organisms. While specific flux was also decreasing as bentonite loading increased, the previous figure (Figure 4.14) showed that increasing specific flux alone (without increasing bentonite loading) had no effect on microbial removal. Thus, the increasing deposition layer appears to be the primary factor for cause of improved microbial removal.

The same trend of improved microbial removal with the accumulation of a deposition layer at pilot scale was discussed above with respect to Table 4.36. At pilot scale, however, the solids loading provided by the natural feedwaters was low compared to that achieved using bentonite addition at bench scale. Thus, the impacts of deposition layer formation on microbial removal were not as pronounced at pilot compared to bench scale. Impacts of deposition layer formation on microbial removal were most consistent only for the poorest-quality feedwater in terms of particulate and organic material (secondary effluent at Scottsdale).

Combining bench- and pilot-scale results, Figure 4.78 illustrates the impact of bentonite (or TSS) mass loading on net MS2 bacteriophage removal (compared to clean water conditions). A clear trend of increasing MS2 bacteriophage removal with increasing solids loading can be observed. Bench- and pilot-scale data points fall along the same trend line, with pilot-scale results at the lower end of the solids loading scale and bench-scale results at the higher solids loadings.

However, no clear relationship is apparent when the bench- and pilot-scale microbial removal results are plotted as a function of specific flux (Figure 4.79) or specific flux decline (Figure 4.80) resulting from deposition layer formation. The bench-scale data form a discernible trend with MS2 bacteriophage removal increasing as specific flux decline (J_S/J_{S0}) decreases in Figure 4.79, and as specific flux decreases in Figure 4.80. However, these trends occur (as observed in Figure 4.80) because decreases in specific flux were achieved at bench scale by increasing the solids loading. Solids loading was relatively low from the natural feedwaters at pilot scale. Thus, resulting declines in specific flux were also relatively low; as such, pilot results do not form consistent trends with the bench-scale results.

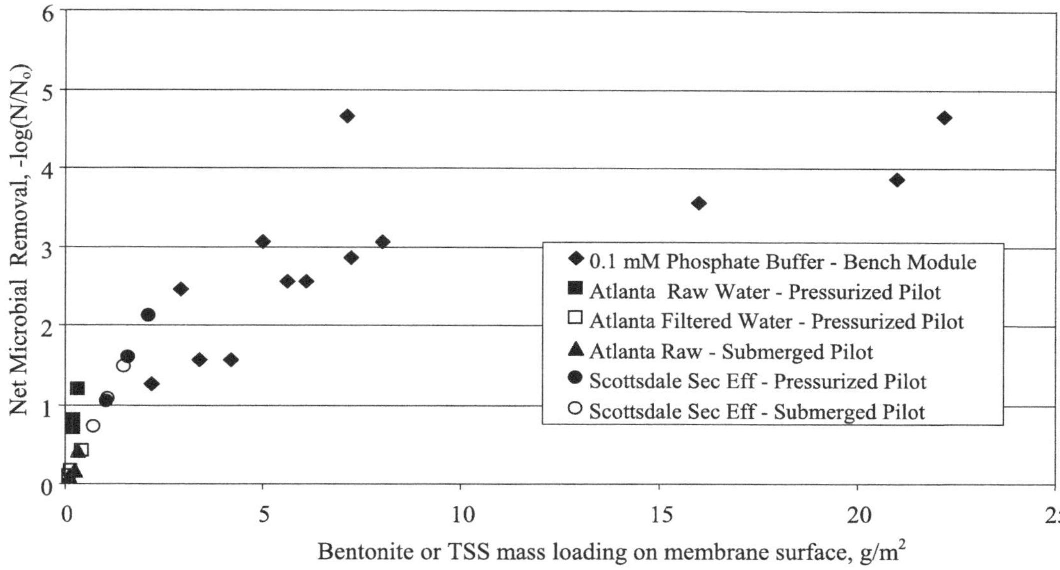

Figure 4.78 Comparative impact of bentonite loading at bench scale and TSS loading at pilot scale on removal of MS2 bacteriophage using PVDF 0.1 μm membrane

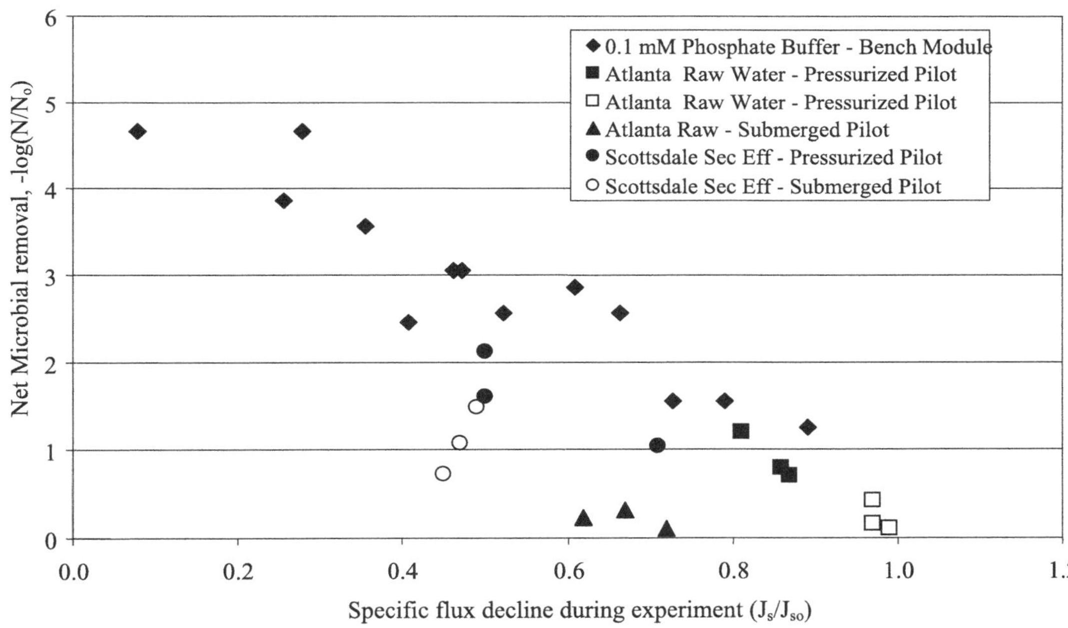

Figure 4.79 Comparative impact of bench- and pilot-scale specific flux decline on removal of MS2 bacteriophage using PVDF 0.1 μm membrane

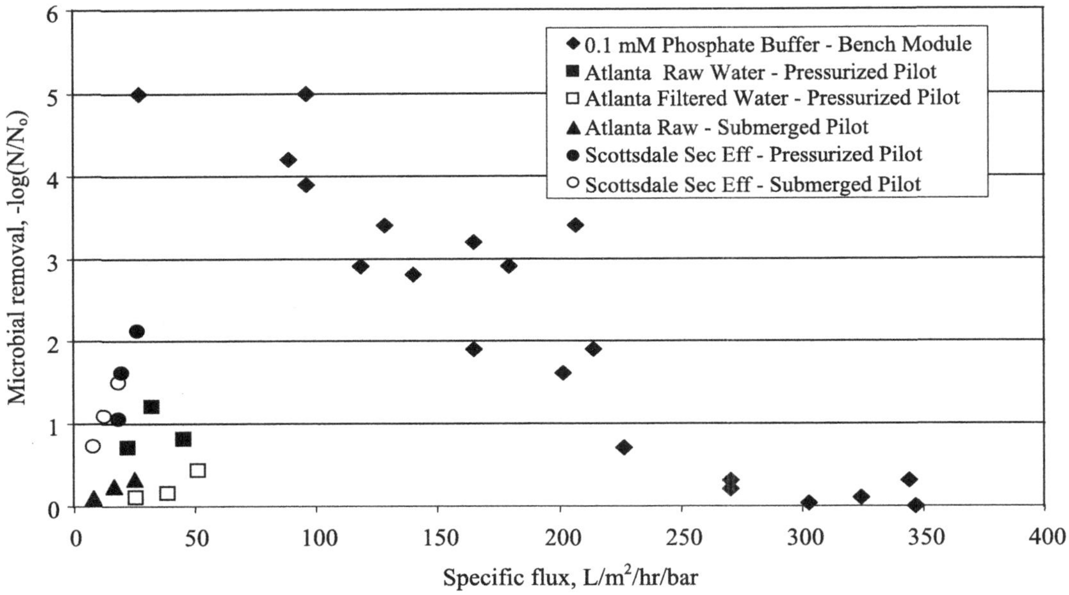

Figure 4.80 Comparative impact of bench- and pilot-scale specific flux on removal of MS2 bacteriophage using PVDF 0.1 μm membrane

COMPARISON OF PILOT- AND FULL-SCALE RESULTS

Scale-Up of Hydraulic Performance and Microbial Removal From Pilot to Full

A pressurized PP membrane pilot unit was operated in parallel with a full-scale pressurized PP membrane system at a similar flux for nine weeks. Identical PP membranes (same membrane material, manufacturer and manufacturer-reported pore size) were used in the pilot and full-scale modules, with the exception that two of the membranes were brand new and one had been in service in the full-scale plant for approximately five years. One of the brand-new PP membrane modules was installed in the pilot plant operated in parallel with the full-scale plant. The other brand-new membrane module was installed in the full-scale plant in a rack containing 89 other PP membrane modules that had been in service in the full-scale plant for approximately five years.

The specific flux of the new PP membrane modules started above 172 $L/m^2/hr/bar$ at 20°C and decreased over the initial two weeks of use on the natural feedwater to stabilize between 74 and 98 $L/m^2/hr/bar$ at 20°C (3.0 and 4.0 gfd/psi at 20°C). The decrease was similar for the two brand-new modules operated at pilot and full scale. The difference in specific flux between the new and mature membranes remained under 25% for the remainder of the testing. After the initial decrease and stabilization, the specific flux observed at pilot scale remained slightly higher than that observed at full scale during the continuous operation but followed the same trend. Chemical cleanings allowed an effective recovery of the specific flux for both pilot- and full-scale membrane modules. These results were discussed with respect to Figure 4.70. Thus, the hydraulic performance of the bench- and pilot-scale membrane modules was comparable.

Comparing the microbial removal performance of the PSN, FSN and FSM membranes, clean water results (see Figure 4.71) indicated that the PSN and FSN membranes provided minimal removal of both MS2 and PRD1 bacteriophage. The FSM membrane provided higher removal of PRD1 bacteriophage (2.0 logs for FSM compared to 0.2 logs or less for PSN and FSN), as well as very slightly higher removal of MS2 bacteriophage (0.4 logs for FSM compared to 0.2 logs or less for PSN and FSN). Thus, the key difference in performance between the tested membranes was not pilot- vs. full-scale membranes, but rather was mature and irreversibly fouled vs. brand new membranes. Similar trends were observed in the 55-day test using natural feedwater for both pilot- and full-scale membrane modules as discussed regarding Figures 4.72 and 4.73. Very little difference in removal of MS2 bacteriophage was observed between pilot- and full-scale membranes. However, the FSM membrane provided up to 3.2 logs of additional PRD1 bacteriophage removal, compared to a removal range of 0.0 to 1.3 logs for both the PSN and FSN membranes.

The full-scale testing showed that a membrane pilot unit can mirror the hydraulic behavior of a full-scale system, and also document satisfactorily the performance of a full-scale membrane system from a microbial removal perspective, provided that the membranes are comparably mature and irreversibly fouled. For brand-new membranes, the pilot and full-scale tests provided comparable hydraulic and microbial removal performance. The key differences in performance were observed between brand-new vs. mature and irreversibly fouled membranes, rather than between pilot-scale vs. full-scale membranes.

Impact of Irreversible Fouling on Microbial Removal

Specific flux and microbial removal data from the side-by-side testing at San Jose illustrates that the impact of irreversible fouling on microbial removal was comparable in both full-scale and pilot-scale membrane systems. Figure 4.70 illustrates the difference between specific flux of the brand new membrane installed on the pressurized pilot unit (PSN) and the full-scale membrane rack containing 89 mature membranes (in use at the full-scale plant for five years) and one brand-new membrane. The specific flux of the full-scale rack was much lower (60 $L/m^2/hr/bar$ at 20°C) than that of the new membrane (145 $L/m^2/hr/bar$ at 20°C) when the side-by-side testing was initiated. Although the additional removal of MS2 bacteriophage (compared to clean water conditions) was minimal for both the pilot- and full-scale PP membranes (Figure 4.72), the additional removal of PRD1 bacteriophage ranged from 0.0 to 1.3 logs for the two new modules and reached 3.2 logs for the mature and irreversibly fouled module (Figure 4.73). Specific flux for the two new membrane modules was higher than that of the mature module at three of the four sampling times before the chemical cleaning. Thus, microbial removal and hydraulic performance relative to maturity and irreversible fouling showed a consistent trend in the pilot- and full-scale membrane modules.

CHAPTER 5
SUMMARY AND APPLICATION TO UTILITIES

SUMMARY OF PROJECT RESULTS

The overall goal of this project was to develop a systematic performance testing protocol and specification for microfiltration and ultrafiltration membranes with respect to removal of viral and submicron bacterial pathogens. Bench-, pilot- and full-scale studies were performed to reach the overall goal. Specific objectives included conducting performance testing at bench scale with a variety of membrane materials and configurations and appropriate viral and submicron bacterial species or surrogates seeded into clean feedwaters; and then performing pilot-scale testing to verify the bench-scale results and performing full-scale testing to validate the pilot-scale data at one utility. A peer-reviewed protocol and performance specification based on data developed at bench scale was also developed for this project. Key issues included the use of microbial surrogates for enteric viruses, and the impact of membrane characteristics, membrane operation and water quality variables on microbial removal.

Bench-Scale Testing

Bench-scale testing was performed on membranes fabricated from a variety of membrane materials and manufacturer-reported MWCOs or pore sizes. Hollow fiber bench testing was performed according to the protocol defined in Chapter 3, Bench-Scale Protocol Development. Membrane feedwater consisted of laboratory deionized, buffered water. Challenge organisms included MS2 and PRD1 bacteriophage, as well as a variety of animal viruses such as poliovirus and hepatitis A virus. The membranes were also challenged with different bacteria and non-biological microspheres.

Surrogates

Microbial Surrogates. MS2 bacteriophage is traditionally employed in water treatment studies as a surrogate for enteric viruses of actual public health concern. In bench-scale testing, a variety of membranes were challenged simultaneously with MS2 and PRD1 bacteriophage as well as enteric viruses including poliovirus 1, feline calicivirus and hepatitis A virus. Correlation coefficients of MS2 bacteriophage removal with PV and HAV removals were greater than 0.90 under the conditions tested. Fewer data points were collected for FCV and correlation was influenced by the sensitivity limit of the FCV assay. Statistical analysis indicated that mean differences in matched MS2 and enteric virus pairs were not statistically different than zero. These results indicate that MS2 bacteriophage is a good surrogate for the enteric viruses PV and HAV under the conditions studied.

PRD1 bacteriophage were generally removed to a greater degree than were the enteric viruses studied. Statistical analyses indicated that mean differences in matched PRD1 and enteric virus pairs were statistically different than zero. These results indicate that PRD1 is not a good surrogate since it was removed to a greater extent than the other viruses. However, PRD1 could be used in conjunction with MS2 to better define the range of microorganisms that could be removed by a particular low-pressure membrane.

Non-Biological Surrogates. The use of fluorescent microspheres was investigated in bench-scale testing. A number of analytical tools were investigated to quantify microsphere concentrations in the membrane feedwater and filtrate, and fluorescence spectrophotometry was selected as the method for this study. Microspheres with 26- and 67-nanometer diameters were evaluated as potential surrogates for MS2 (0.024-/0.026-µm diameter) and PRD1 (0.070-µm diameter) bacteriophage. The microspheres were not found to be good surrogates for the bacteriophage under the range of experimental conditions employed.

Membrane Characteristics

Manufacturer-Reported MWCO or Pore Size. A key objective of this study was to evaluate the validity of using manufacturer-reported MWCOs or pore sizes for estimating microbial removal. Bench-scale testing was performed using five hollow-fiber modified-polysulfone membranes that varied only in manufacturer-reported pore size (same manufacturer, same membrane material), ranging from 10 kD to 0.2 µm. While the "loosest" membrane (0.2 µm) provided the poorest removal of MS2 and PRD1 bacteriophage and poliovirus, the 100- and 300-kD membranes provided consistently higher removal of the three organisms than did the "tightest" (10 kD) membrane. Further, the specific flux of the membranes was not necessarily consistent with their manufacturer-reported MWCOs or pore sizes. While the 10-kD membrane had the lowest specific flux and the 0.2-µm membrane had the highest, the specific flux of the 300-kD membrane was lower than that of the 100 kD membranes. Thus, manufacturer-reported MWCO or pore size, and specific flux, were not necessarily good predictors of microbial removal under the conditions tested.

Membrane Operation

Transmembrane Flux. Bench-scale microbial challenge testing was performed using three different membranes (differing in manufacturer-reported pore sizes or MWCOs and membrane materials) at transmembrane pressures ranging from 0.1 to 2.1 bar (1.5 to 30 psi), which produced specific fluxes ranging from 35 to almost 500 L/m^2/hr/bar (20 to 300 gfd/psi). Within the range tested, increasing specific flux did not have any effect on removal of MS2 or PRD1 bacteriophage or hepatitis A virus.

Decreased Specific Flux in the Presence of a Deposition Layer. Bench-scale microbial challenge testing was performed using bentonite addition to form a deposition layer on the membrane surface. Some testing was conducted using the PVDF 0.1 µm membrane at a constant bentonite loading to evaluate the impact of pressure/flux on microbial removal. Increasing the transmembrane pressure (increasing the flux) did not increase removal of either MS2 bacteriophage or poliovirus. Thus, compaction of the deposition layer did not lead to improved removal of bacteriophage or virus under the conditions tested.

Membrane Integrity. Membrane integrity is a key issue for effective microbial removal by membrane filtration. Bench-scale experiments were performed to evaluate the impact of membrane compromising on removal of MS2 and PRD1 bacteriophage. Flat sheet membranes were compromised by puncturing with a needle to produce one hole on the order of 100 µm in diameter, which was equivalent to approximately 0.005% of the membrane surface area. The effect on microbial removal was substantial, with one hole capable of reducing removal of MS2 and PRD1 bacteriophage to near zero for membranes that had provided up to 3.0 logs of bacteriophage removal in an uncompromised condition. It should be noted, however, that in comparison to hollow fibers, this breach is equivalent to one out of every 10 fibers being impaired. Thus, one should be careful in any extrapolation of results from one membrane geometry to the other.

Water Quality

Impact of pH. Bench testing was performed to evaluate the impact of pH on microbial removal. A number of membranes of varying materials were tested for pH effects, including the PVDF 0.1 μm, the PVDF-hpi, PVDF-hpo, RC-hpi and PES-hpo. Microorganisms tested included MS2 and PRD1 bacteriophage, as well as poliovirus. All of the bench testing indicated that pH had no impact on microbial removal within a pH range comprising the typical water treatment pH levels (5 to 9).

Impact of Ionic Strength. The impact of ionic strength on microbial removal was evaluated in bench-scale experiments using sodium chloride addition to achieve conductivities ranging from 64 to 54,100 μs/cm. As the concentration of salt increased, the removal of the bacteriophage also increased. However, the effect was observed at concentrations normally associated with a brackish groundwater as opposed to a typical surface water. There was a clear trend between ionic strength and microbial removal.

Presence of a Deposition Layer. The presence of a deposition layer or hydraulically-reversible fouling on microbial removal was evaluated in experiments employing bentonite addition for deposition layer formation. The bench-scale membranes were challenged with bacteriophage and virus after the bentonite suspension had been added to the feedwater to allow solids accumulation on the membrane surface. Bentonite addition had the effect of decreasing specific flux by decreasing filtrate production at a given pressure. In experiments investigating the impact of specific flux at a given bentonite mass loading, results indicated that changing trans-membrane flux did not impact microbial removal. These findings suggest that there was no relationship between specific flux and microbial removal under the conditions tested.

In other experiments, the bentonite mass loading was increased and the effects on microbial removal were quantified. As bentonite loading increased, removal of MS2 and PRD1 bacteriophage improved, as did removal of poliovirus. Specific flux declined as bentonite loading was increased. However, the previous set of experiments had demonstrated that the decline in specific flux was not itself associated with improved microbial removal; rather, the increase in the deposition layer (increased bentonite mass loading) was the cause of improved microbial removal.

Impact of Microbial Loading. Bench-scale testing was performed to determine if high microbial loading could cause "leakage" of the microorganisms into the filtrate. Membranes of varying manufacturer-reported MWCOs or pore sizes were exposed to widely varying MS2 bacteriophage loading rates. Results indicated that MS2 bacteriophage were removed to the sensitivity limits of the assay by all membranes at concentrations up to 10^5 pfu/cm^2. However, as MS2 concentrations increased to 10^7 pfu/cm^2, MS2 bacteriophage removal by one of the membranes (mPS 10 kD) remained constant and then actually declined at a concentration of 10^8 pfu/cm^2. However, it is important to note that the effect of microbial loading on microbial removal was membrane-specific. It should also be noted that the microbial loading in these experiments were extremely high and are very unlikely to encountered in drinking water treatment applications.

Pilot-Scale Testing

Pilot-scale testing was performed in the United States at Atlanta and Scottsdale, and in the United Kingdom at London. Both pressurized and submerged membrane systems were employed, and a variety of membrane materials and manufacturer-reported MWCOs or pore sizes (0.1-μm PVDF in the US, and 0.035-μm PP and 150-kD PES/PVP in the UK). Membrane feedwaters included raw and filtered water from municipal drinking water treatment plants (Atlanta and London) and secondary effluent from a wastewater treatment plant (Scottsdale). Challenge organisms for the pilot-scale testing included MS2 and PRD1 bacteriophage as viral surrogates, and *Pseudomonas diminuta* bacteria. Pilot testing included initial determination of microbial removal under clean water conditions, which were the worst-case conditions for microbial removal (clean water, no deposition layer or fouling on the membrane). Following clean water testing at each pilot-testing location, microbial removal was evaluated under conditions of deposition layer formation or hydraulically-reversible fouling, followed by a longer-term evaluation of microbial removal under conditions of hydraulically-irreversible fouling. All pilot tests were performed at a constant flux.

Membrane Characteristics

Manufacturer-Reported MWCO or Pore Size. Pilot tests at Atlanta and Scottsdale employed PVDF membranes with a manufacturer-reported pore size of 0.1 μm. Pilot tests at London employed PP and PES/PVP membranes with manufacturer-reported MWCOs or pore sizes of 0.035 μm and 150 kD, respectively. Pilot testing results using clean water (dechlorinated tap water) at the various locations indicated that removal of MS2 and PRD1 bacteriophage corresponded to the manufacturer-reported MWCO or pore size of the membrane. The membranes with the smallest manufacturer-reported MWCO (150 kD) removed the highest levels of both MS2 and PRD1 bacteriophage, and the membranes with the largest manufacturer-reported pore size (0.1 μm) removed the lowest levels of MS2 and PRD1 bacteriophage.

However, in agreement with bench-testing results, manufacturer-reported MWCO or pore size was not always predictive of bacteriophage removal. For example, a membrane with a manufacturer-reported pore size of 0.1 μm provided a variety of levels of MS2 bacteriophage removal, from 0.0 to 1.5 logs. The bench- and pilot-scale results suggest that size exclusion was not the only mechanism producing microbial removal; or alternatively, a single membrane may have a wide size distribution of MWCOs or pore sizes so that there are pores in the membrane both larger and smaller than the manufacturer-reported nominal MWCO or pore size.

The two membranes with the smallest manufacturer-reported MWCOs or pore sizes (0.035 μm and 150 kD) provided such high levels of removal of MS2 and PRD1 bacteriophage under clean water conditions that any improvement in microbial removal due to deposition layer formation or membrane fouling was not consistently discernible under the conditions tested.

Membrane Maturity and Irreversible Fouling. The PVDF membranes used in the submerged pilot unit were brand-new at the time pilot testing for this study was initiated. The PVDF membranes used in the pressurized pilot unit had been in use on natural water supplies previously. Microbial removal and specific flux data from clean water testing of the brand-new membrane used in the submerged pilot unit initially at Atlanta and then approximately six months later at Scottsdale demonstrate the impacts of membrane maturity and irreversible fouling, e.g., fouling that cannot be removed by backwashing or by chemical cleaning. Removal of both MS2 and PRD1 bacteriophage increased by approximately 1.0 log and specific flux at 20°C decreased by more than 30% as the membrane matured and became irreversibly fouled to some extent during approximately six months of use on natural feedwaters.

Pressurized vs. Submerged Membranes. Pilot testing results from Atlanta and Scottsdale using PVDF membranes of identical membrane material and manufacturer-reported pore size (0.1 μm) indicated that pressurized versus submerged membrane systems affected microbial removal and membrane operation under clean water conditions. Under clean water conditions, the submerged PVDF membrane provided some removal of MS2 bacteriophage at both pilot-testing locations (0.2 to 0.4 log at Atlanta and 1.1 to 1.4 logs at Scottsdale), while the pressurized PVDF membrane did not remove any appreciable levels of MS2 bacteriophage at either of those locations. The submerged PVDF membrane consistently experienced greater specific flux decline than the pressurized PVDF membrane in deposition layer formation experiments although it was operated at lower fluxes. However, the same trend was not observed consistently in longer-term membrane fouling experiments.

Water Quality

Deposition Layer Formation. In deposition layer formation pilot testing, the membranes were operated for 4 hours without backwashing. MS2 and PRD1 bacteriophage were seeded into the feedwater and their concentrations in the membrane feedwater and filtrate were quantified. Microbial removal was calculated and then normalized to account for the baseline removal under clean water conditions. The additional microbial removal (the difference between removal during the deposition layer formation experiments and the clean water experiments) was that attributable to the accumulation of a deposition layer on the membrane surface.

In general, deposition layer formation (as characterized by the decline in specific flux over the 4-hour experiment) varied according to the feedwater and membrane employed. For example, when raw water was used as the feedwater, specific flux decline reached a maximum of 38% in Atlanta. When using the submerged PVDF membrane in London, during testing with the membrane having the smallest manufacturer-reported MWCO (150 kD), the maximum specific flux decline was 77% with raw water. When filtered water from the water treatment plant was used as the feedwater, specific flux decline reached a maximum of only 16% after 4 hours regardless of the membrane employed. When secondary wastewater effluent was used as the feedwater, specific flux decline reached a maximum of 55% during the 4 hours of membrane operation.

The presence of a deposition layer led to improved microbial removal (compared to clean water conditions) on a consistent basis only for the secondary effluent feedwater, which had a substantially higher organic and particulate content than the raw or filtered feedwaters tested. In Scottsdale using secondary effluent as feedwater, the presence of a deposition layer increased PRD1 bacteriophage removal by a maximum of 3.1 logs, and increased MS2 bacteriophage removal by up to 1.6 logs compared to clean water conditions. In Atlanta, deposition layer formation increased PRD1 removal by a maximum of 1.8 logs and MS2 bacteriophage removal by a maximum of 1.3 logs in Atlanta with raw water; however, these increases were not consistent over all experimental conditions. In London, removal of both MS2 and PRD1 bacteriophage under clean water conditions was sufficiently high that consistent improvement with deposition layer formation was not observed.

Increases in PRD1 bacteriophage removal with deposition layer formation compared to clean water conditions were consistently higher than those of MS2 bacteriophage removal in pilot testing at Atlanta and Scottsdale. This observation is consistent with the relative sizes of the two microorganisms, with PRD1 bacteriophage having the larger diameter (0.065 μm) compared to MS2 bacteriophage (0.025 μm).

Membrane Fouling. The impact of membrane fouling on microbial removal was evaluated in pilot-scale experiments performed over several days to two weeks. The membranes were operated at a constant flux and at a backwash interval selected to allow membrane operation for an extended period before reaching the maximum operating transmembrane pressure when chemical cleaning would be required. MS2 and PRD1 bacteriophage were seeded into the membrane feedwater and their removal was quantified.

Membrane fouling had a much greater impact on microbial removal than did deposition layer formation under the conditions tested at all pilot plant locations. Declines in specific flux were greater during the longer-term membrane fouling experiments compared to the deposition layer formation experiments, and removal of PRD1 bacteriophage was consistently improved with membrane fouling compared to clean water conditions. Removal of the smaller-diameter MS2 bacteriophage was not as consistently improved with membrane fouling.

Flux declines ranged from 30 to 90% for all but two of the 11 fouling experiments performed for this study. In all tests at Atlanta, membrane fouling led to increased removal of PRD1 bacteriophage, regardless of feedwater (raw or filtered) and membrane type (pressurized or submerged PVDF membranes). At the end of the experiment, maximum additional PRD1 bacteriophage removal compared to clean water conditions ranged from 1.6 to 4.9 logs at Atlanta. Additional removal of MS2 bacteriophage was not as consistent as that of PRD1, ranging from 0.0 to 0.9 logs for raw and filtered water for the submerged PVDF membrane and the pressurized PVDF membrane using filtered water, but reached 5.0 logs for the pressurized PVDF membrane using raw water. In Scottsdale pilot testing using secondary effluent as feedwater, maximum flux decline over the 2-week experiments ranged from 30% for the submerged PVDF membrane, producing a maximum net PRD1 bacteriophage removal of 4.1 logs, to 90% for the pressurized PVDF membrane, which produced a maximum net PRD1 removal of 5.4 logs and a maximum net MS2 bacteriophage removal of 4.4 logs.

In US pilot testing, the pressurized PVDF membrane was the only one of the two membranes tested for which fouling produced high removals of MS2 bacteriophage removal, using raw water at Atlanta (up to 5.0 logs at a maximum specific flux decline of 58%) and using secondary effluent at Scottsdale (up to 4.4 logs at a maximum specific flux decline of 90%). This membrane appears to have been irreversibly fouled to some extent since after terminating the experimental run, the specific flux recovered from a 90% decline to only a 60% decline upon chemical cleaning after the two-week test.

Microbial removal with membrane fouling was less consistent in testing at London. While specific flux declines were substantial for most of the experiments (up to 50%), the additional removal of PRD1 did not exhibit a consistent trend for either of the two membranes (submerged POL or pressurized PES/PVP) or feedwater types (raw or filtered) in London testing. For MS2, additional removal reached 3.7 logs for the pressurized PES/PVP membrane, but the trend of increasing microbial removal with increasing flux decline was inconsistent.

Full-Scale Testing

Field testing comparing pilot-scale and full-scale membrane performance was performed in San Jose, California. At the 5-mgd full-scale membrane plant, raw surface water is subjected to pressure-driven membrane filtration using polypropylene membranes with a manufacturer-reported pore size of 0.2 μm. The full-scale membrane system comprises six membrane racks, each containing 90 membrane modules. The pressurized pilot unit was operated side-by-side with the full-scale membrane plant. Three PP membrane modules were evaluated. One brand-new module was installed and operated in the pilot plant. One brand-new module was installed in one rack of the full-scale plant. The third module evaluated was one of the modules that had been in service in the full-scale plant since 1994. In order to evaluate the microbial removal performance of the three modules, each was installed in the pilot unit where microbial seeding could be performed. Upon completion of the microbial seeding, each of the two modules from the full-scale plant was replaced in the full-scale membrane rack to resume operation until the next microbial challenge, when each was again removed from the full-scale rack and installed in the pilot plant.

Side-by-side testing indicated that the hydraulic performance of the pilot-scale membrane module was comparable to that of the full-scale membrane rack. Specific flux was considerably higher in the pilot-scale membrane module initially, because it was brand new and was being compared to the full-scale membrane rack that contained 89 mature and somewhat irreversibly fouled membrane modules and one brand-new module. However, after approximately two weeks of operation using the natural raw water as feedwater, the specific flux of the pilot membrane module stabilized no more than approximately 25% higher than that of the full-scale membrane rack. Flux recovery following chemical cleaning was consistent between the pilot- and full-scale membrane modules.

Using both clean water (dechlorinated tap water) and raw water as feedwater, microbial removal performance of the brand-new membranes from the pilot- and full-scale membrane systems was comparable. Using clean water, neither of the two brand-new membrane modules, whether from the pilot- or the full-scale system, removed MS2 or PRD1 bacteriophage to any extent (maximum removal of 0.2 log). However, the mature membrane module removed 0.4 logs of MS2 bacteriophage and 2.2 logs of PRD1 bacteriophage. Thus, although the membranes were similar in every way except their history of use, PRD1 bacteriophage removal by the mature membrane was substantially higher that that of the brand-new membranes.

When the membranes were operated side-by-side using natural feedwater, again there was little difference in MS2 bacteriophage removal between the pilot- vs. full-scale membrane modules, or between the brand-new vs. mature and irreversibly fouled membranes. For PRD1 bacteriophage, there was little difference in removal between the brand-new pilot- and full-scale membrane modules. However, removal of PRD1 by the mature and irreversibly fouled membrane module was higher by 1.0 to 2.0 logs than PRD1 bacteriophage removal by either of the two brand-new membranes.

Thus, the key finding of the full-scale testing was that there were substantial differences in PRD1 removal performance between brand-new vs. mature and irreversibly fouled membranes. New membranes produced substantially lower removal of PRD1 bacteriophage than mature and irreversibly fouled membranes at similar specific fluxes. There was little removal of MS2 by any of the membranes, whether brand-new or mature/irreversibly fouled, or whether pilot scale or full scale.

OVERALL CONCLUSIONS

Overall findings of the bench-, pilot- and full-scale testing conducted for this project are:

- MS2 bacteriophage is a good surrogate for the enteric viruses studied (poliovirus, hepatitis A virus and feline calicivirus). Use of MS2 bacteriophage as a viral surrogate was included in the bench-scale membrane testing protocol developed during this project.
- MS2 bacteriophage removal was generally lower than that of PRD1 bacteriophage; MS2 bacteriophage is therefore the worst-case surrogate. Use of PRD1 bacteriophage was included in the bench-scale membrane testing protocol developed during this project.
- Microspheres are not a good surrogate for viruses and were not included in the bench-scale membrane testing protocol developed during this project.
- Manufacturer-reported MWCO or pore size is not always predictive of microbial removal performance.
- Specific flux did not affect microbial removal performance under the conditions tested.
- Membrane integrity was crucial for microbial removal. Even a hole comprising only 0.005% of the membrane area dramatically reduced microbial removal.
- pH did not affect microbial removal within a range typical of water treatment applications (pH 5 to 9).
- Ionic strength did not affect microbial removal within a range typical of water treatment applications.
- The natural raw waters tested in this study had sufficiently high particulate and organic quality that deposition layer formation did not have consistent effects on microbial removal. Deposition layer formation consistently improved microbial removal only when secondary wastewater effluent or other high-TSS feedwater (TSS of approximately 5 mg/L or higher) was used as the membrane feedwater.
- Membrane fouling, whether reversible (chemical cleaning resulted in recovery specific flux) or irreversible (chemical cleaning did not produce full recovery of specific flux), caused much greater improvements in microbial removal than did deposition layer formation.
- Membrane maturity (i.e., the unrecoverable decline in flux observed when a brand-new membrane is placed into operation for some length of time) caused improvements in microbial removal.
- Microbial removal performance was comparable at bench- and pilot-scale, whether under clean water conditions or in the presence of a deposition layer.
- Hydraulic and microbial removal performance was comparable in pilot-scale and full-scale membrane operation.

APPLICATION TO UTILITIES

Bench-Scale Membrane Testing Protocol

A major outcome of this project is the bench-scale membrane testing protocol that was presented in Chapter 3. This protocol was developed as a standardized, cost-effective means of characterizing membranes for their capability to remove microbial contaminants. Such a protocol is needed because microbial removal by membrane systems typically is evaluated through pilot studies. However, there are numerous reasons that bench-scale microbial challenge testing of membranes may be more feasible than pilot testing. Costs of pilot testing are most often much higher than bench-scale testing, primarily due to equipment and staffing. Pilot plants are typically located at water treatment facilities, so it is often not possible to use challenge organisms due to risk concerns. Further, there is typically more scatter in microbial removal data from pilot studies since it is difficult to control accurately experimental conditions. The well controlled conditions of bench scale allow for the generation of more accurate and precise experimental results.

This study showed that the presence of a deposition layer or fouling on the membrane can improve microbial removal; thus, it is critical to sample a clean membrane as soon as it is placed in to service. This is much simpler to accomplish at bench scale than pilot scale. Further, the use of clean water, such as laboratory deionized water, which does not contain solid or organic material that can cause deposition layer formation or fouling, is also a particular challenge to the membrane. Thus, the bench-scale protocol allows the membrane to be challenged under actual worst-case conditions; that is, a brand-new, clean membrane immediately after being placed into service using clean water. Microbial challenges performed under these conditions are an actual representation of the membrane's capability to remove microbial contaminants that is not confounded by deposition or fouling. It should be noted, however, that bench-scale testing does not take into account full-scale module construction or system imperfections that can cause microbial passage to the filtrate.

The protocol employs a hollow fiber geometry for the bench-scale testing unit because most low pressure membrane applications for drinking water involve hollow fiber modules. However, the protocol can be employed for other membrane geometries with little modification. The intent of the protocol is to provide a widely accepted, standardized methodology with which to characterize membranes from a microbial perspective.

The protocol was developed for scientific rigor, but also for ease of implementation by a qualified laboratory. The protocol will produce a cost-effective, standardized, rigorous characterization of microbial removal by membranes that will be a useful screening tool for any water provider considering membrane treatment of their water supplies. By following the protocol, data will be developed to provide utilities, engineers and regulators with the necessary information to make informed decisions about current and future membrane products that are being (or will be) employed for treatment of water supplies.

Bench-Scale Protocol Is Conservative

As discussed above, the bench-scale protocol developed for this study characterizes microbial removal by membranes under worst-case conditions of clean membrane and clean water. This is the only standardized means of assessing microbial removal for a given membrane because the membrane begins to mature as soon as it is operated on natural waters. Membrane maturity is the unrecoverable decline in flux observed when a brand-new membrane is placed into operation for some length of time, due in large part to the accumulation of NOM from natural waters on the surface of or within the membrane. This study showed the importance of membrane fouling and maturity for microbial removal, with improved microbial removal produced by a fouled or mature membrane compared to a clean or new membrane. Thus, the bench-scale protocol presents the most conservative conditions under which to characterize the membrane's microbial removal capabilities. In practice, the actual microbial removal achieved by the membrane would be higher due to membrane maturity and fouling.

Deposition Layer Improves Microbial Removal Only for Very Poor Water Qualities

This study demonstrated that the presence of a deposition layer on the membrane led to improved microbial removal. A deposition layer is hydraulically-reversible fouling comprised of accumulated solid and organic material from natural feedwaters. It is removed by backwashing. However, this study showed that membrane feedwaters typical of those used for membrane treatment (e.g., relatively low in particulate and organic content) did not produce deposition layers that reliably improved microbial removal, even during a 4-hour filtration cycle. Typical filtration cycles, or the time between backwashes, was on the order of 30 minutes or less for all of the membranes used in pilot testing for this study. Thus, a relatively poor-quality feedwater is necessary in order to reliably achieve the microbial-removal benefits of a deposition layer. In this study, the only feedwater that produced a deposition layer over a 4-hour filtration cycle that reliably improved microbial removal was the secondary clarified wastewater. From a practical perspective, data generated in this study suggest that deposition layer should not be relied upon to provide consistent additional virus removal.

Pore Size Is Not Absolute

This study demonstrated repeatedly in bench and pilot testing that the manufacturer-reported pore size or MWCO is not predictive of microbial removal. Many factors other than simple size exclusion affect microbial removal by membranes using natural water sources. The bench-scale membrane testing protocol developed for this study is a means of making a valid characterization of microbial removal by low pressure membranes in order to provide more accurate and complete information than manufacturer-reported pore size MWCO.

Successful Scale-Up From Bench to Pilot to Full Scale

This study showed that microbial removal results produced by the bench-scale membrane testing protocol were representative of pilot-scale microbial removal performance for a given membrane. The study further demonstrated in side-by-side testing that pilot-scale microbial removal was representative of that achieved at full scale. The study also demonstrated that flux profiles observed at pilot scale were comparable to those observed at full scale.

MEMBRANE CLASSIFICATION

The overall goal of this project was to develop a performance testing protocol and specification for low pressure membranes with respect to removal of viral and submicron bacterial pathogens. An important task to meet this goal was to develop a peer-reviewed protocol and performance specification; this was followed by a second task which was to conduct performance testing at bench scale with a variety of different membranes using a clean water feed suspension of organisms. For pilot studies, there are various "accepted" protocols that are used for microbial challenge testing of membranes (NSF and USEPA, 2002; USEPA, 2003b). However, there is currently no generally accepted methodology for microbial removal bench-scale testing of membranes used for drinking water. Most protocols have come out of the pharmaceutical and microelectronics industries, with no two protocols alike. The protocol that was developed in this study is presented in Chapter 3. As noted, it was designed for the application of sound science combined with ease of implementation by most qualified laboratories. To promote its use, it was developed with the idea that it would be employed under the Environmental Technology Verification (ETV) Program managed by the NSF for the USEPA. As such, it has been adopted as the bench-scale option to perform microbial challenge studies for membranes (NSF and USEPA, 2002). Additionally, the protocol was written with the intent to be employed as a bench-scale option described in the *Membrane Filtration Guidance Manual* (GM; USEPA 2003b) to the LT2ESWTR. The draft GM currently describes in detail how full-scale membrane modules should be challenged with microorganisms. However, it does not stipulate at any length the manner in which bench microbial challenge studies should be preformed. Rather, it states, "For the purposes of consistency, it is recommended that manufacturers or independent testing agencies that opt to subject a product line to challenge testing using small-scale modules utilize a protocol that has been accepted by a wide range of stakeholders. Such a protocol has been proposed for use under the National Sanitation Foundation (NSF) Environmental Technology Verification (ETV)." The latter is the protocol developed under this study.

Using the protocol developed, hollow fiber membranes varying in manufacturer-reported MWCO/pore size as well as membrane materials were evaluated for removal of selected microorganisms. Table 5.1 presents a proposed virus and bacterial classification (rating) for the membranes studied. The table shows the average removal of MS2 and PRD1 bacteriophage as well as their respective lower confidence interval value around the average at P = 0.01. Also shown is a proposed classification of the membranes (rating) in terms of virus and bacterial removal. The scale employed is based on using half log values associated with the lower value of the confidence interval. The virus rating employs the removal of MS2 bacteriophage; this organism was chosen because of its small size (24 μm) and good correlation with removal of the enteric viruses studied. For example, the mPS 10 kD membrane had a lower confidence interval value of 3.8 log for removal of MS2; thus, its designation would be v-3.5.

For bacteria, removal of *P. diminuta* would be employed as the chosen organism by which to determine the bacterial rating of the membrane. This organism was selected based on its small size (0.6 to 1 μm). However, if no challenge studies with the bacteria were conducted and the membrane removed either MS2 or PRD1 bacteriophage to the RSL of the organism, the lower value of the confidence interval for the removal of the virus would be employed for the rating. If both PRD1 and MS2 data sets are generated, the lower value of the confidence interval for the bacteriophage that was removed to the greatest extent would be employed.

For example, for the PVDF 0.1 μm membrane, challenge studies with *P. diminuta* were conducted and the lower confidence interval value for the removal of the organism was 6.8 log. Thus, the bacterial rating for this membrane would be b-6.5. However, for the mPS 300 kD membrane, no bacterial challenge studies were performed. Consequently, the MS2 bacteriophage data would be employed and the bacterial rating would assume the same rating as the virus and become b-5.0.

Table 5.1
Viral and bacterial ratings of hollow fiber membranes employed in bench-scale study using Low-Pressure Membrane Testing Unit

		MS2 bacteriophage		PRD1 bacteriophage			
Module designation	MWCO or pore size*	Average log removal	Lower confidence interval value at P = .01	Average log removal	Lower confidence interval value at P = .01	Virus rating	Bacterial rating
mPs 10 kD	10,000 daltons	3.8	3.8	5.0	4.6	v-3.5	nc
mPS 100 kD	100,000 daltons	>5.7	5.7	>5.9	5.7	v-5.5	b-5.5
mPs 300 kD	300,000 daltons	>5.5	5.5	>5.2	4.8	v-5.0	b-5.0
PVDF 0.035 μm	0.035 μm	3.8	3.7	4.5	4.2	v-3.5	nc
PVDF 0.1 μm	0.1 μm	0.34	0.29	0.53	0.49	v-0	6.5†
mPS 0.2 μm	0.2 μm	0.45	0.22	nd	nd	v-0	nc

* As reported by the manufacturer
† Bacterial rating based upon the lower confidence interval removal value obtained from challenge studies of the membrane with *P. diminuta* (6.8 log-not shown on this table)
nd = no data
nc = not classified

APPENDIX A

MATERIALS EMPLOYED FOR FABRICATION OF BENCH-SCALE LOW-PRESSURE MEMBRANE TEST UNIT FOR CHARACTERIZATION OF MEMBRANES USING MICROBIAL PROFILING

(Note: if needed, contact NSF for potential source of materials.)

- Pressure vessel with vacuum closure (2-gallon volumes)
- Glass-filled nylon instant tube ("plug and play") fitting, male pipe adapters ¼", ¼"
- Glass-filled nylon instant tube ("plug and play") fitting, male 90 elbow pipe adapters ¼", ¼"
- Glass-filled nylon instant tube ("plug and play") fitting, male branch tee pipe adapters ¼", ¼", ¼"
- Glass-filled nylon instant tube ("plug and play") fitting, male run tee pipe adapters ¼", ¼", ¼"
- Glass-filled nylon instant tube ("plug and play") fitting, coupling ¼", ref 5779k14
- Glass-filled nylon instant tube ("plug and play") fitting, 90 elbow ¼", ref 5779k24
- Glass-filled nylon instant tube ("plug and play") fitting, tee ¼", ref 5779k34
- Cement: All-purpose cement for PVC, ABS, CPVC and reference 30821
- Nylon tubing: named "Nylon 6 tubing" and order in ¼" OD, ref 5173K9
- PVC threaded pipe fitting – schedule 80 – dark gray, reducing hex bushing, NPT male ¾" × NPT female ½", ref 4596k414
- Miniature chrome-plated brass ball valves, female ¼", female ¼", wedge handle, ref 4912k47
- Teflon thread sealant tape, ½" width, ref 4591k12

APPENDIX B

QUALITY ASSURANCE/QUALITY CONTROL FOR BENCH-SCALE PROTOCOL FOR CHARACTERIZATION OF MEMBRANES USING MICROBIAL PROFILING

Introduction

Quality assurance and quality control of the operation of the membrane equipment and the measured water quality parameters shall be maintained during the laboratory testing program.

Operational and Low-Pressure Membrane Testing Unit QA/QC

Before the testing of each manufacturer's modules, on-line pressure gauges shall be checked with secondary gauges to confirm that the read-out matches the actual measurement. Unit tubing and connections shall be inspected weekly to verify that they are in good condition. Replacement of these materials shall be made as necessary.

Analytical Methods

The analytical methods utilized in this study for feedwaters are described below.

pH

Analyses for pH shall be performed according to Standard Method 4500-H^+ (APHA et al., 1998). A 3-point calibration of the pH meter used in this study shall be performed once per day when the instrument is in use. Certified pH buffers in the expected range shall be used. The pH probe shall be stored in the appropriate solution defined in the instrument manual.

Temperature

Readings for temperature shall be conducted in accordance with Standard Method 2550 (APHA et al., 1998). The thermometer shall have a scale marked for every 0.1°C, as a minimum, and shall be calibrated biweekly against a precision thermometer certified by the National Institute of Standards and Technology. (A thermometer having a range of -1°C to +51°C, subdivided in 0.1°C increments, would be appropriate for this work.)

Turbidity

Turbidity analyses shall be performed according to Standard Method 2130 (APHA et al., 1998) using a bench-top turbidimeter. All glassware used for turbidity measurements shall be cleaned and handled using lint-free tissues to prevent scratching. Sample vials shall be stored inverted to prevent deposits from forming on the bottom surface of the cell. Grab samples shall be analyzed using a bench-top turbidimeter. Information on calibration, verification of calibration, sampling and analysis can be found in the *ETV Protocol for Equipment Testing for Physical Removal of Microbiological and Particulate Contaminants* (NSF and USEPA, 2002).

Particle Counting

Bench-top particle counters shall be used to measure particle concentrations in the feed-water. Laser light scattering or light blocking instruments are recommended for particle counting; however, other types of counters such as Coulter counters or Elzone counters may be considered.

The following particle size ranges shall be monitored by bench-top analytical instruments during the membrane characterization testing:

- 2-3 μm
- 3-5 μm
- 5-7 μm
- 7-10 μm
- 10-15 μm
- >15 μm

Information on calibration, verification of calibration, maintenance of the particle counters, particle free water, sampling and analysis can be found in the *ETV Protocol for Equipment Testing for Physical Removal of Microbiological and Particulate Contaminants* (NSF and USEPA, 2002).

Conductivity

This parameter shall be measured according to Standard Method 2510B (APHA et al., 1998).

Total Organic Carbon (TOC) /Dissolved Organic Carbon (DOC)

TOC/DOC shall be analyzed according to Standard Method 5310B or 5310C (APHA et al., 1998).

Chlorine Preparation for Membrane Cleaning

The stock solution shall be prepared by adding an estimated volume of 6% reagent-grade NaOCl into a 500-mL, chlorine demand-free bottle containing an estimated amount of organic-free water. The concentration of the chlorine solution will be as recommended by the manufacturer. Refer to Standard Method 4500-Cl F (APHA et al., 1998) for the preparation method of DPD indicator, FAS standard and buffer solution. Residual free chlorine measurements will be conducted according to Standard Method 4500-Cl G, DPD Colorimetric Method. (APHA et al., 1998)

Bacteriophage

Bacteriophage MS2 and PRD1 shall be enumerated according to NWRI and AwwaRF (2000). Because of the importance of these organisms in characterizing the membrane at bench scale, detailed methods for MS2 are provided below and shall be followed.

MS2 Bacteriophage Soft Agar Overlay Method and MS2 Stock Preparation. The bacteriophage MS2 - ATCC 15597-B1 shall be employed in all studies. The *Escherichia coli* C-3000 – ATCC 15597 shall be employed as the host bacterium, with the bacterial growth media being tryptic soy broth (TSB) - DIFCO 0370-15-5, or the equivalent.

Tryptic Soy Agar (Bottom Agar Petri Plates 100 × 15mm) Preparation. The media is rehydrated according to label directions. A magnetic stir bar is placed into the dehydration flask, and the media is brought to a near boil to dissolve the agar. The flask and contents are then sterilized by autoclaving for 15 minutes after which it is cooled in a water bath to between 45 to 50°C. Plates are poured using approximately 12-15 mL per plate. Enough agar is added to cover two-thirds of the area of the plate. After pouring one plate, the lid is replaced on the dish and gently swirled so that the agar covers the entire bottom of the plate. Plates are allowed to remain motionless until the agar hardens (usually 10-15 minutes). Plates are stored at 4°C up to 30 days.

Tryptic Soy Agar Overlay Tubes. The media is rehydrated according to label directions. A magnetic stir bar is placed into the rehydration flask, and the media is brought to a near boil to dissolve the agar. The media is sterilized for 15 minutes and then pipetted aseptically into 15-mL tubes (3 mL per tube) or pipetted into the tubes (3 mL per tube), which are then capped loosely and sterilized for 15 minutes. The caps are tightened after cooling. Overlay tubes are stored at 4°C for 30 days.

Preparation of High-Titer MS2 Bacteriophage. To propagate the MS2 bacteriophage, a bacterial host slant of *E. coli* (ATCC #15597) is washed with 3 mL of sterile tryptic soy broth (TSB). The total 3 mL is then transferred to a 1-liter flask containing 200 mL of sterile TSB and incubated at 37°C for approximately 3 hours. At this time, the flask is removed from the incubator and 2 mL of bacteriophage stock (ATCC #15597-B1) is added and then the flask is placed back in the incubator for an additional 4 hours. Then, 0.02 g of lysozyme and 6 mL of 0.2 M sterile EDTA are added to the flask which is shaken for an additional 30 minutes (not included in Debi's method). The bacteriophage/bacteria suspension is poured into four 50-mL centrifuge tubes and centrifuge at 4,000 times gravity for 15 minutes.

During preparation of high-titer MS2 bacteriophage stocks, there is a potential for aggregate formation. To reduce aggregates, the MS2 stock is filtered through sequentially smaller (0.45 micron, 0.22 micron and 0.1 micron [01. micron filtration not included in Debi's method]), low protein-binding filters. To reduce MS2 binding to the filter, each filter is pretreated by filtering 10 mL of 0.1% Tween 80 followed by 10 mL reagent grade non-chlorinated water. MS2 stock preparations are filtered through these pretreated filters with careful attention focused on amount of pressure/vacuum applied to prevent membrane filter failure. Multiple filters may be necessary to filter the entire MS2 stock solution.

The bacteriophage stock is then titered to determine its concentration and stored at 4°C for up to 4 weeks.

Preparation of Host Culture. The host culture is started the day before the assay is to be performed. Using a sterile swab, a small amount of *E. coli* host (ATCC 15597) is removed from an agar slant and placed into a sterile tube containing 3 mL of tryptic soy broth and grown overnight at 37°C for 24 hours. The next day, 1 mL of the overnight culture is pipetted into 50 mL of tryptic soy broth in a 250-mL Erlenmeyer flask or the equivalent. The culture is then placed in a 37°C incubator for 4 hours, after which it is removed from incubator and placed on ice until used.

Soft Agar Overlay Method for Bacteriophage. Bacteriophage in bench-scale low-pressure membrane samples are enumerated by the addition of the sample to soft or overlay agar along with a liquid culture of bacteria (host) in the log phase of growth. Overlay tubes are melted in a boiling water bath or autoclaved for 5 minutes and placed in a 49°C water bath until used. The bottom of the petri plates are labeled with the identification of sample to be analyzed. Then 0.1 to 1 mL of the 4-hour host culture (which is in log phase of growth) is pipetted into a pre-warmed overlay tube along with 0.1 to1 mL of the sample to be analyzed. The tube is mixed by rapidly rolling between the analyst's palms and poured onto a TSB plate. The sample is spread evenly over the surface of the plate by gently and quickly swirling the plate. The plate, which solidifies within 30 seconds, is then inverted and incubated at 37°C for 24 ± 2 hours. During the incubation time, the host bacteria forms a confluent lawn over the surface of the petri plate. The petri plate is incubated at 37°C for 24 hours. During the incubation period, the phage particles that are present in the sample attach to and infiltrate the bacterial host cells. The bacteriophage replicate within the bacterial cells and reach a concentration that lyse (burst) the bacterium. The destruction of the bacterial cells that make up the confluent lawn result in clear areas known as plaques. The concentration of bacteriophage originally present in the sample are determined by visually counting the clear areas, which are reported number of plaque forming units per mL (PFU/mL).

Pseudomonas diminuta

This organism was enumerated according to ASTM Method F838-83 (1993).

Escherichia coli

This organism was enumerated according APHA, AWWA, and WPCF (1998).

REFERENCES

APHA, AWWA, and WEF (American Public Health Association, American Water Works Association, and Water Environment Federation). 1998. *Standard Methods for the Examination of Water and Wastewater.* 20th ed. Washington, D.C.: APHA.

ASTM International. 1993. *Standard Test Method for Determining Bacterial Retention of Membrane Filters Utilized for Liquid Filtration.* Method F838-83 (re-approved 1993). West Conshohocken, Pa.: ASTM International.

Baker, R.W. 2000. *Membrane Technology and Applications.* New York: McGraw-Hill.

Cho, J. 1998. Natural Organic Matter (NOM) Rejection by, and Flux Decline of, Nanofiltration (NF) and Ultrafiltration (UF) Membranes. Ph.D. diss., University of Colorado at Boulder, Colo.

Fane, A.G. 1994. An Overview of the Use of Microfiltration for Drinking Water and Waste Water Treatment. In *Proc. of Microfiltration for Water Treatment Symposium.*

Herath, G., K. Yamamoto, and T. Urase. 1998. Mechanism of Bacterial and Viral Transport Through Microfiltration Membranes. *Water Science and Technology,* 38(4–5):489–496.

Herath, G., K. Yamamoto, and T. Urase. 1999. Removal of Viruses by Microfiltration Membranes at Different Solution Environments. *Water Science and Technology,* 40(4–5):331–338.

Hunter, P.R. 1997. *Waterborne Disease: Epidemiology and Ecology.* New York: John Wiley & Sons.

Izard, D., C. Ferragut, F. Gavini, K. Kersters, J. De Ley, and H. Leclerc. 1981. *Klebsiella terrigena*—a New Species From Soil and Water. *International Jour. of Systematic Bacteriology,* 31:116–117.

Jacangelo, J.G., S.A. Adham, and J.M. Laine. 1995. Mechanism of *Cryptosporidium, Giardia* and MS2 Virus Removal by Microfiltration and Ultrafiltration. *Jour. AWWA,* 87(9):107.

Jacangelo, J.G., E.M. Aieta, K.E. Carns, E.W. Cummings, and J. Mallevialle. 1989. Assessment of Hollow Fiber Ultrafiltration for Particulate Removal in Drinking Water. *Jour. AWWA,* 81(11):68.

Jacangelo, J.G., J.M. Laine, K.E. Carns, E.W. Cummings, and J. Mallevialle. 1991. Low-Pressure Membrane Filtration for Removing *Giardia* and Microbial Indicators. *Jour. AWWA,* 83(9):97.

Joklik, W.K., H.P. Willet, D.B. Amos, and C.M. Wilfert, eds. 1992. *Zinsser Microbiology.* 20th ed. Norwalk, Conn.: Appleton & Lange.

Lide, D.R., ed. 2004. *Handbook of Chemistry and Physics.* 85th ed. Boca Raton, Fla.: CRC Press.

Lytle, C.D., and L.B. Routson. 1995. Minimized Virus Binding for Tests of Barrier Materials. *Applied and Environmental Microbiology,* 61(2) 643–649.

Madaeni, S.S. 1998. The Application of Membrane Technology for Water Disinfection. *Water Research,* 33(2):301–308.

Madaeni, S.S., A.G. Fane, and G.S. Grohmann. 1995. Virus Removal From Water and Wastewater Using Membranes. *Jour. of Membrane Science,* 102:65–75.

McGahey, C.L. 1994. Mechanisms of Virus Capture From Aqueous Suspension by a Polypropylene, Microporous, Hollow-Fiber Membrane Filter. Ph.D. diss., The Johns Hopkins University, Baltimore, Md.

NSF International and USEPA (U.S. Environmental Protection Agency). 2002. *ETV Protocol for Equipment Verification Testing for Physical Removal of Microbiological and Particulate Contaminants*. Ann Arbor, Mich.: NSF International.

NWRI and AwwaRF (National Water Research Institute and Awwa Research Foundation). 2000. *Ultraviolet Disinfection Guidelines for Drinking Water and Water Reuse*. No. 2674. Denver, Colo.: AwwaRF.

Olivieri, V.P., G.A. Willinghan, J.C. Vickers, and C. McGahey. 1991. Continuous Microfiltration of Secondary Wastewater Effluent. In *Proc. of AWWA Membrane Technology in the Water Industry Conference*.

Schijven, J.F., and M. Hassanizadeh. 2002. Modeling of Virus Transport and Removal in the Subsurface. In *Encyclopedia of Environmental Microbiology*, Volume 4. Edited by G. Bitton. New York: John Wiley & Sons.

Shim, Y., H-J. Lee, S. Lee, S-H. Moon, and J. Cho. 2002. Effects of Natural Organic Matter and Ionic Species on Membrane Surface Charge. *Environmental Science and Technology*, 36:3864–3871.

Streeter, V.L., and E.B. Wiley. 1989. *Fluid Mechanics*. 8th ed. New York: McGraw-Hill.

USEPA (U.S. Environmental Protection Agency). 1989. Surface Water Treatment Rule. 40 CFR 141.70-40.

USEPA (U.S. Environmental Protection Agency). 1998. National Primary Drinking Water Regulations: Interim Enhanced Surface Water Treatment Rule; Final Rule, 40 CFR Parts 9, 141 and 142, December 16.

USEPA (U.S. Environmental Protection Agency). 2002. National Primary Drinking Water Regulations: Long Term 1 Enhanced Surface Water Treatment Rule—Final Rule. 40 CFR Parts 9, 141 and 142, January 14.

USEPA (U.S. Environmental Protection Agency). 2003a. National Primary Drinking Water Regulations: Long Term 2 Enhanced Surface Water Treatment Rule—Proposed Rule. 40 CFR Parts 141 and 142. August 11.

USEPA (U.S. Environmental Protection Agency). 2003b. *Membrane Filtration Guidance Manual*. EPA 815-D-03-008. Proposal Draft. June.

Watson, G. 2002. Performance and Integrity Testing of Low Pressure Membranes: Microspheres as Bacteriophage Surrogates, Master's thesis. University of Colorado at Boulder, Colo.

Wright, S. 2002. Relating Complex Solute Mixture Characteristics to Membrane Fouling Using Flow Field Fracitonation. Ph.D. diss., University of Colorado at Boulder, Colo.

Yoo, R.S., D.R. Brown, R.J. Pardini, and G.D. Bentson.1995. Microfiltration: A Case Study. *Jour. AWWA*, 87(3):38–49.

ABBREVIATIONS

Å	angstrom
AFC WTP	Atlanta Fulton County Water Treatment Plant
AFM	atomic force microscopy
ATCC	American Type Culture Collection
AWTW	Advanced Water Treatment Works
AwwaRF	Awwa Research Foundation
BOD	biological oxygen demand
C	Celsius
cfu	colony forming unit
CIP	clean in place
cm	centimeter
COD	chemical oxygen demand
DOC	dissolved organic carbon
ETV	Environmental Technology Verification
FCV	feline calicivirus
FSM	full-scale mature
FSN	full-scale new
g	gram
gal	gallon
gfd	gallon per square foot per day
gfd/psi	gallon per square foot per day per pound per square inch
HAV	hepatitis A virus
HPC	heterotrophic plate count
hpi	hydrophillic
hpo	hydrophobic
hr	hour
HRT	hydraulic residence time
in	inch
J	flux
JHU	The Johns Hopkins University
J_s	specific flux
J_{s20}	specific flux at 20°C
kD	kilodalton

L	liter
L/hr	liter per hour
$L/m^2/h$	liters per square meter per hour
$L/m^2/hr/bar$	liters per square meter per hour per bar
LT1ESWTR	Long-Term 1 Enhanced Surface Water Treatment Rule
LT2ESWTR	Long-Term 2 Enhanced Surface Water Treatment Rule
m	meter
M	molar
m^2	square meter
m^3	cubic meter
MEM	Minimal Essential Medium
MF	microfiltration
mg	milligram
mgd	million gallons per day
mg/L	milligram per liter
µm	micrometer
min	minute
mL	milliliter
mM	millimolar
mPS	modified polysulfone
µS/cm	microsiemens per centimeter
mV	millivolt
MWCO	molecular weight cut-off
nm	nanometer
NOM	natural organic matter
NSF	National Sanitation Foundation
NTU	nephelometric turbidity unit
PBS	phosphate-buffered saline
PDT	pressure decay test
PES	polyethersulfone
pfu	plaque forming unit
PLC	programmable logic controller
PS	polysulfone
psi	pound per square inch
PSN	pilot scale new
P_{tm}	transmembrane pressure
PV	poliovirus
PVC	polyvinyl chloride
PVDF	polyvinylidine fluoride
PVP	polyvinyl pyrrolidone
QA	quality assurance
QC	quality control

RB	roller bottle
RC	regenerated cellulose
RSL	removal sensitivity limit
sec	second
SEM	scanning electron microscopy
SUVA	specific ultraviolet absorbance at 254 nanometers
SWC	Scottsdale Water Campus
SWTR	Surface Water Treatment Rule
TKN	total kjeldahl nitrogen
TOC	total organic carbon
TSB	tryptic soy broth
TSS	total suspended solids
UCB	University of Colorado, Boulder
UF	ultrafiltration
UK	United Kingdom
US	United States
USEPA	United States Environmental Protection Agency
USF	University of South Florida
UV-254	ultraviolet light absorbance at 254 nanometers
WTP	water treatment plant

IWA Publishing's authorised EU representative for General Product Safety Regulations is Diane D'Arras, 15 rue Duret, 75116 Paris, France, e-mail: safety@iwap.co.uk.

Printed and bound by CPI Group (UK) Ltd, Croydon, CR0 4YY
17/09/2026
02218793-0003